CHEMICAL ENGINEERING METHODS AND TECHNOLOGY

AMMONIA

STRUCTURE, BIOSYNTHESIS AND FUNCTIONS

CHEMICAL ENGINEERING METHODS AND TECHNOLOGY

Additional books in this series can be found on Nova's website
under the Series tab.

Additional E-books in this series can be found on Nova's website
under the E-book tab.

BIOCHEMISTRY RESEARCH TRENDS

Additional books in this series can be found on Nova's website
under the Series tab.

Additional E-books in this series can be found on Nova's website
under the E-book tab.

CHEMICAL ENGINEERING METHODS AND TECHNOLOGY

AMMONIA

STRUCTURE, BIOSYNTHESIS AND FUNCTIONS

VICTORIA A. FEKETE
AND
RÉKA L. MOLNÁR
EDITORS

Nova Science Publishers, Inc.
New York

For permission to use material from this book please contact us:
Telephone 631-231-7269; Fax 631-231-8175
Web Site: http://www.novapublishers.com

LIBRARY OF CONGRESS CATALOGING-IN-PUBLICATION DATA

Ammonia : structure, biosynthesis, and functions / editors, Victoria A. Fekete and Rika L. Molnar.
p. cm.
Includes index.
ISBN 978-1-62100-502-5 (hardcover)
1. Ammonia. I. Fekete, Victoria A. II. Molnar, Rika L.
TP223.A46 2011
546'.7112--dc23
2011034487

Published by Nova Science Publishers, Inc. † New York

CONTENTS

PREFACE

Ammonia is a natural and common nitrous agent affecting all vital processes in animal, plant and bacterial cells. In organisms, it is produced by about two hundred enzyme reactions, thus being an essential and harmless metabolite. At high concentrations, ammonia becomes a strong toxin. In this book, the authors present current research in the study of the structure, biosynthesis and functions of ammonia. Topics include the biochemical studies on energy metabolism in animals in acute ammonia intoxication; development of distributed fiber optic sensors of ammonia gas; inhibition of rRNA synthesis by amines and ammonium ions in xenopus embryos; amino acids that play roles in plant adaptation to abiotic stress and the atmospheric concentration of NH3, NO2, HNO3 and SO2 by the passive method compared with corresponding emission inventory.

Chapter 1 - Acute administration of the lethal dose of ammonia results in the rapid death of animals. This review includes data on the role of energy metabolism in ammonia-induced mortality. The studies reviewed here show that acute ammonia intoxication leads to the quick depletion of metabolic substrates such as glycogen, glucose, ketone bodies and ATP, first in the liver and second in the brain in vivo, finished with coma and death. The following effects of acute ammonia intoxication mainly in non-synaptic brain mitochondria will be considered: (1) oxidative phosphorylation, malate-aspartate shuttle, calcium transport and the membrane potential; (2) antioxidative and pro-oxidative enzymes and other parameters of oxidative stress; (3) cytochrome c release, DNA fragmentation, PARP activation, p53 transfer and other markers of neuronal apoptosis. The roles of glutamate NMDA receptors and the nitric oxide system as well as association of mitochondrial, cytosolic and nuclear processes in acute hyperammonemia are briefly discussed.

Chapter 2 - Our contribution starts with a brief overview of the recent state-of-art in the field of 'classical' sensors routinely used in detection of ammonia gas. A short reference is made of a wide practical usage of ammonia gas and its harmful properties stimulating the ever-lasting emphasis on development of spatially continuous and highly sensitive sensors. Consecutive summary of alternative approaches taking advantage of utilization of optical fibers in place of the sensing element is then followed by a detailed theoretical treatment of the fiber optic distributed system employing the optical time domain reflectometry (OTDR) technique. The derived model is used in computer simulations, results of which are compared with the experimental data obtained in tests of real sensing fibers. Suggestions are then asserted concerning the promising directions of further development of the sensor system.

Chapter 3 - In *Xenopus* embryogenesis, transcription of rRNA genes begins shortly after midblastula stage (or at the transition called MBT), and its activity increases greatly thereafter (Shiokawa et al., 1981a,b). We were interested in the control mechanism of this MBT-associated rRNA gene activation, and tried to find out substances which control rRNA gene expression. We examined the blastula cell conditioned medium, the blastula homogenate, and its acid-soluble fraction. After various tries and errors, we eventually reached the conclusion that weak bases inhibit quite selectively the synthesis of rRNA but not tRNA and mRNA and the inhibition takes place at the transcriptional level. In the present article, we first summarize our studies on initiation of rRNA gene expression during MBT, or the transition from the cleavage stage to the post-blastular stages in *Xenopus* developing embryos. We then summarize in some details our efforts performed to find out factors that control rRNA gene expression. Then, finally we describe our quite unexpected discovery that weak bases such as amines and ammonium ion selectively inhibit rRNA gene expression in *Xenopus* embryonic cells. We analyzed acid-extractable substances in *Xenopus* cleavage stage embryos and found that a larger amount of ammonia is present in pre-blastula stage embryos than in the post-blastular stage embryos. We also found that replacement of Na^+ with choline$^+$ in the culture medium completely abolishes the inhibition of rRNA gene expression. We, therefore, conclude that ammonium ion is one of the components that regulate rRNA gene expression in *Xenopus* embryogenesis, acting probably by inducing a slight increase in the intracellular pH.

Chapter 4 - Plants absorb nitrogen as nitrate or ammonium ions from the soil, and the nitrogen is assimilated into the amino acids. Under the environmental stress conditions, plant assimilation of nitrogen and the metabolic pathway in amino acid biosynthesis can be affected. In drought and salinity conditions, amino acids, such as proline, accumulate and function as osmolytes that affect osmotic adjustment in plant cells. Proline synthesis also affects the biogenesis of reactive oxygen species (ROS). Phenylalanine is synthesized with glutamate and converted to *trans*-cinnamic acid by phenylalanine ammonia-lyase (PAL), which catalyzes the first reaction in phenylpropanoid metabolism. The expression of a variety of genes that function in the metabolic pathway to increase stress tolerance is upregulated in plant cells. In this chapter, we present nitrogen assimilation under stress conditions and focus on the transcriptome and metabolome studies in regulatory networks in plant abiotic stress tolerance.

Chapter 5 - Annual emission map for NH_3, NO_X and SO_2 in Japan was shown according to the EAGrid2000-Japan emission database. The median emission of NH_3, NO_X and SO_2 in the 10 X 10 km grid was 0.37, 0.69, 0.078 ton/km^2/y, respectively, while that at the 30 sites was 2.4, 22, 3.3 ton/km^2/y, respectively. Monthly emission for NH_3 showed apparent seasonal trends, being high in summer and low in winter. In the case of NO_X and SO_2, the emission was slightly high in winter and low in summer and constant through the year, respectively. Atmospheric concentration of NH_3, NO_2, HNO_3 and SO_2 by the passive method was compared with corresponding emission inventory. Average concentrations of HNO_3, SO_2, NH_3 and NO_2 were 5.6-39.7, 11-146, 34-175 and 93-1191 nmol/m^3, respectively. The emission inventory flux of SO_2, NO_X and NH_3 was investigated within 1km^2, 100km^2 and 1300km^2 zones including the sampling sites. The correlation for NH_3 was significant in the three emission zones. The correlations for NO_2, HNO_3 and SO_2 were also significant, although with some exception. As the emission inventory included rather high stack (more than 25m) facilities combustion sources, the correlation probably was good in large sphere rather than the small sphere. Monthly average concentration of NH_3, NO_2, HNO_3 and SO_2 was shown at the sites where performed relatively long term survey during FY2003-2006. Monthly concentration of NH_3 was high from July to November, while monthly emission of NH_3 was high from June to September (summer) and low from December to March (winter). The temporal trend of NO_2 concentration was high in winter and low in summer similar to that of NO_X emission. In contrast, the trend of HNO_3 concentration was high in summer and low in winter, reverse to that of NO_X emission. There was not particular seasonal trend of SO_2 concentration, where SO_2 emission had also not seasonal variation.

Chapter 6 - Ammonia flux has been monitored continuously since July 2008 over semi-natural grassland at the Hungarian NitroEurope site '*Bugac-puszta*'on the Great Hungarian Plain. Results presented here are based on the data obtained from July to September, i.e., during the vegetation period. The instrument used for ammonia concentration gradient measurement was a novel diode laser based photoacoustic device combined with preconcentration sampling (WaSul-Flux), developed at the University of Szeged. Ammonia concentration measurements were performed at three different levels (0.5 m, 1.3 m and 3 m), on a cc. 30-minute accumulation interval. The three inlets were moved automatically to the same level (1.3 m) twice a week by a remote controlled automated system to check the precision of the measurement. The turbulent flux of ammonia was calculated using the similarity theory based on eddy covariance data of momentum, heat, water vapor and carbon dioxide fluxes (provided by a CSAT3 sonic anemometer and a LICOR-7500 open path CO_2/H_2O sensor), in view of the friction velocity (u_*) and the Monin-Obukhov length scale (L). Sensitivity analyses of ammonia flux calculation as (i) calculation of ammonia gradient, (ii) choice of universal function and (iii) application of different gradient and profile techniques, have been investigated. The diurnal variation of the ammonia concentration and flux has also been investigated. During the studied period the net daytime emission and nocturnal deposition were observed with large deviation exceeding the average flux values both during day and night. The daily mean ammonia concentrations were compared to data measured at the Hungarian background air quality monitoring station (K-puszta) ~20 km far from the Bugac-puszta site, and fairly good agreement was found between the two datasets.

In: Ammonia: Structure, Biosynthesis... ISBN: 978-1-62100-502-5
Editors: V.A. Fekete, et al, pp. 1-32

Chapter 1

ENERGY METABOLISM IN ACUTE AMMONIA INTOXICATION

Elena A. Kosenko and Yury G. Kaminsky*
Institute of Theoretical and Experimental Biophysics,
Russian Academy of Scienses, Pushchino, Russia
and Pushchino State University, Pushchino, Russia.

ABBREVIATIONS

ROS, reactive oxygen species;
NMDA-R, NMDA receptor;
PARP, poly(ADP-ribose) polymerase;
MAO, monoamine oxidase;
SOD, superoxide dismutase.
A short vertical arrows indicates a direction of a parameter change.

ABSTRACT

Acute administration of the lethal dose of ammonia results in the rapid death of animals. This review includes data on the role of energy metabolism in ammonia-induced mortality. The studies reviewed here

* E-mail: kaminsky@iteb.ru

show that acute ammonia intoxication leads to the quick depletion of metabolic substrates such as glycogen, glucose, ketone bodies and ATP, first in the liver and second in the brain in vivo, finished with coma and death. The following effects of acute ammonia intoxication mainly in non-synaptic brain mitochondria will be considered: (1) oxidative phosphorylation, malate-aspartate shuttle, calcium transport and the membrane potential; (2) antioxidative and pro-oxidative enzymes and other parameters of oxidative stress; (3) cytochrome c release, DNA fragmentation, PARP activation, p53 transfer and other markers of neuronal apoptosis. The roles of glutamate NMDA receptors and the nitric oxide system as well as association of mitochondrial, cytosolic and nuclear processes in acute hyperammonemia are briefly discussed.

I. Introduction

Ammonia is a natural and common nitrous agent affecting someways all vital processes in animal, plant and bacterial cells. In the organism, it is produced by about two hundred of enzyme reactions, thus being the essential and harmless metabolite. However, at high concentrations, ammonia becomes a strong toxin. In this article, results of biochemical studies on energy metabolism in animals in acute ammonia intoxication are reviewed.

Biochemistry and physiology of ammonia in vitro have been well described in the classic review by Cooper and Plum [16] and are not topic of this work. Biochemical processes related to chronic effects of ammonia on organisms as well as amonia toxicity for isolated organ systems and cell cultures will not be condsidered, too.

II. Ammonia Metabolism Disturbance and Hyperammonemia

Low tissue ammonia levels are supported by the urea cycle which is completely only in the liver of all mammal species. Within other tissues ammonia is removed by reductive amination of 2-oxoglutarate in the glutamate dehydogenase reaction and amidation of glutamate to form glutamine in the glutamine synthetase reaction. Three- and morefold increase in blood ammonia levels is latently considered as ammonia metabolism disturbance. Hyperammonemia arises in many human and animal pathologies such as acute and chronic liver and kidney deficiencies, hepatoencephalopathy, Reye syndrome, Alzheimer's disease, alcohol intoxication, organ transplantation and other conditions.

Hairy hyperammonemia is associated with genetic errors such as deficiency of one or more enzymes of the urea cycle. Total screening allowed to show up one child with an inherited disorder of ammonia metabolism per fourteen thousand of neonates [86].

III. Ammonia Toxicity

As a chemical, ammonia is known of early 18th century, however its biochemistry began to investigate much later. At late 19th century, the Russian scientist Ivan Pavlov in cooperation with Polish Marcelius Nencki and colleagues nave found that dogs with portocaval anastomosis died with convulsions and hyperammonemia soon after consumption of beef. Thus, ammonia toxicity for animals and the role of the liver in detoxication were discovered for the first time [29, 48, 75]. Nearly 80 years ago, serious mental disorders in patients with ascitic cirrhosis given with ammonium chloride as a diuretic were described [101, 102]. Ammonia toxicity for people was demonstrated in such unusual way.

Cellular mechanisms underlying ammonia toxicity leading to damage to nervous cells are not wholly elucidated. As seizures and coma are highlights of hyperammonemia, it is commonly believed that ammonia is neurotoxin, not hepatotoxin. Thereby biological effects of ammonia on the liver was studied insufficiently and the literature data are scant, contradictive and require rectification.

IV. Biochemical Changes in Acute Ammonia Intoxication: Sequence of Metabolic Events

Toxicity of large doses of ammonia is showed by emergence of convulsion episodes in first 5-8 min, hyperventilation, clonic convulsions 5-7 min later, coma and rapid lethal outcome [47, 62, 96]. Therefore biochemical studies on ammonia toxicity are usually performed during 15-20 min following an injection of the acute dose.

Figure 1 shows changes of the key metabolite levels in rat blood, liver and brains simultaneously at 0, 5, 10 and 15 min after ammonium chloride injection [47]. These results have not been presented before in the literature available. Glucose, acetoacetate and 3-hydroxybutyrate were increasingly depleted in the blood and liver in 10 min, before emergence of convulsions. Brain 2-oxoglutarate concentration does not change through 15 min [32, 47], disproving the well-known Krebs-cycle depletion theory of hepatic coma [9].

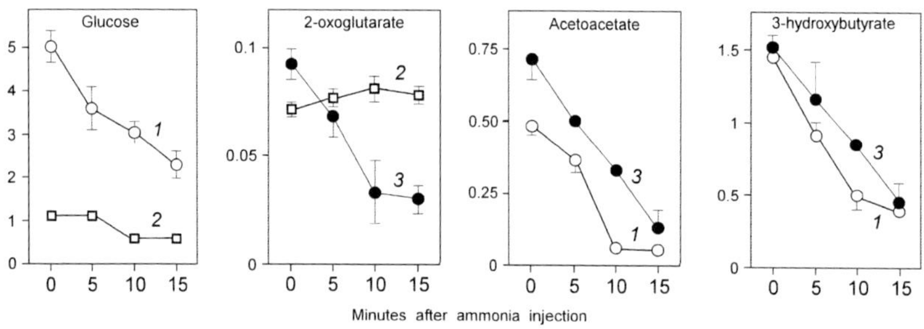

Figure 1. The time course of glucose, 2-oxoglutarate, acetoacetate and 3-hydroxybutyrate in the blood, liver and brains from rats fasting for 24h during first 15 min after an injection of ammonium chloride (by [47]). 1. blood; 2, brain; 3, liver. Metabolite concentrations are expressed as mmol per liter or kg of tissues.

In the Hawkins et al. [32] study, brain glucose utilization was measured after a single intravenous injection of [2-^{14}C]glucose to 48h-fasted rats and was calculated to be increased by 29% 5 min after ammonium acetate injection. What source of surplus (5-6 mM) blood glucose could be involved?

Brain ATP does not change during first 2.5 min [96], 5 min [32] or 10 min [47], but decreases as much as 6-fold proximately before animal death [47]. Rats which did not develop spontaneous periodic clonictonic convulsions recovered fully at 30 min after ammonium acetate injection, however the basilar ATP concentration was 30% decreased [96].

Because the brain cannot synthesize glucose, it critically depends on a continuous supply of glucose from the circulating blood and hence from gluconeogenesis, the process proceeding principally in the liver [38, 76]. The rate of glucose production from endogenous substrates in hyperammonemic rat liver homogenate is 5 times lower than that in the control preparation [47]. It was first and is only evidence for the strong inhibition of gluconeogenesis *ex vivo* with ammonia administered. It is commonly believed that, when gluconeogenesis is depressed under hypoglycemic conditions, brain metabolism commutes glucose oxidation to the oxidation of ketone bodies, the latter is produced by the liver, too [76]. Depletion of blood and liver ketone bodies during convulsions (Figure 1) indicates that ketogenesis is severe suppressed in acute hyperammonemia. The comatose state, induced with the lethal dose of ammonium chloride, is accompanied by pronounced hypoglycemia and almost disappearance of blood acetoacetate and 3-hydroxybutyrate, by decreases in liver 2-oxoglutarate, pyruvate, lactate and ketone bodies, not precedes these metabolic alterations. Hence, both ketosis and acidosis do not prompt the initiation of coma, but are important consequences of ammonia toxicity [47]. Above changes of energy metabolism in the blood, liver and brain suggest that ammonia intoxication is characterized by at least two stages. At first, disturbances in the liver take place. Before emergence of the ammonia coma, liver metabolic processes related to energy providing of all organ, tissues and cells, namely gluconeogenesis and then ketogenesis are inhibited. Second, brain metabolism disturbs some later, only in the comatous state. Ammonia plays a key role in the pathogenesis of hepatic encephalopathy, which manifests as a neuropsychiatric syndrome accompanying acute and chronic liver failure [31].

V. Brain Energy Metabolism in Acute Ammonia Intoxication

Impaired bioenergetics seems to be one of proposed mechanisms of ammonia toxicity for the cell. Administration of the large dose of ammonium acetate to animals results in disturbance of brain adenine nucleotide metabolism as soon as 11-15 min after injection [39, 40, 51]. In acute hyperammonemia, all characteristics of the cellular energy potential such as the ATP concentration, total adenine nucleotide pool, adenylate energy charge and phosphorylation potential decrease while those of depletion of energy stores such as ADP, AMP, inorganic phosphate levels and ATPase activity increase. This pattern may reflect disturbances in mitochondrial energy generation, oxidative phosphorylation.

VI. Mitochondrial Energy Metabolism

Activity of the citric acid cycle and the rate of ATP production by oxidative phosphorylation are inhibited, oxidative stress increases and lactate is accumulated as a result of ammonia toxicity. Short literature reviews on mitochondrial dysfunction in acute hyperammonemia were published recently [24, 25, 88]. Further, it will be shown in some details how ammonia affects *in vivo* most important functional properties of mitochondria: oxidative phosphorylation, the malate-aspartate shuttle, enzymes and metabolites, calcium transport, and antioxidant status.

VI.1. *In Vitro* and Ex Vivo Effects of Ammonia on Oxidative Phosphorylation

Effects of ammonia on mitochondrial oxidative metabolism is under study from 1961. McKhann and Tower [72] were first who discovered that ammonia is an inhibitor of the mitochondrial respiration *in vitro*. In their experiments, 10 mM ammonium chloride inhibited the phosphorylating oxidation of pyruvate and 2-oxoglutarate by isolated cat cerebral cortex mitochondria and did not affect succinate and glutamate oxidation, while 15 mM and 40 mM ammonium chloride were required to inhibit glutamate and succinate respiration, respectively. All ammonium chloride, ammonium acetate and ammonium sulfate at 1.25-2.5 mM inhibited succinatc plus acetate oxidation by rat liver mitochondria [43]. Other workers reported reduced that 14 mM ammonium chloride inhibited oxidation of pyruvate and 2-oxoglutarate by rat liver mitochondria but did not affect succinate and malate oxidation [109].

In rat brain homogenate ammonium chloride at 5-20 mM inhibits phosphorylating and uncoupled respiration with succinate, 3-hydroxybutyrate, pyruvate plus malate, and glutamate plus malate without effects on the state 4 respiration (our unpublished observations). Maximum effects of ammonia were 25-35% at 10 mM. Acute inection of ammonium acetate to rats influened alike [54]. These results showed unequivocally that ammonium ion is an inhibitor of rat brain mitochondrial oxidation of all respiratory substrates and does not uncouple oxidative phosphorylation, both in vitro and in vivo.

VI.2. Ex Vivo Effects of Ammonia on Malate-Aspartate Shuttle

For cytosolic NADH to be oxidized through the respiratory chain, a transfer of reducing equivalents from the cytosol into mitochondria must occur. The malate-aspartate shuttle is a major route in the brain for the transfer [14, 27]. A scheine of the malate-aspartate shuttle is shown in Figure 2. The shuttle is a closed cycle involving the transport of malate and glutamate into the mitochondrion in exchange for intramitochondrial 2-oxoglutarate and aspartate, respectively. The activities of malate dehydrogenase and aspartate aminotransferase in both the mitochondrion and cytosol are also involved in the shuttle. Cytosolic NADH is oxidized to NAD by oxaloacetate in the reaction catalyzed by cytosolic malate dehydrogenase. The resulting malate enters the mitochondrion (through a malate-oxoglutarate antiporter) to be converted back to oxaloacetate plus NADH by an intramitochondrial malate dehydrogenase. As the inner mitochondrial membrane is hardly permeable to oxaloacetate [28], this substrate, when formed intramitochondrially, cannot efflux into the cytosol directly; however, it does so, after conversion to aspartate, by transamination with intramitochondrial glutamate via a mitochondrial aspartate aminotransferase. Aspartate leaves the mitochondrion through a glutamate-aspartate antiporter. In the cytosol, aspartate is transaminated by cytosolic aspartate aminotransferase, replenishing cytosolic pools of glutamate and oxaloacetate. Another cycle begins, resulting in the removal of one NADH molecule from cytosol and yielding one NADH molecule in the mitochondrion. The malate-aspartate shuttle partially reconstituted with brain mitochondria from hyperammonemic rats is inhibited by 20% as compared to that in brain mitochondria from cotltrol aninlals [54].

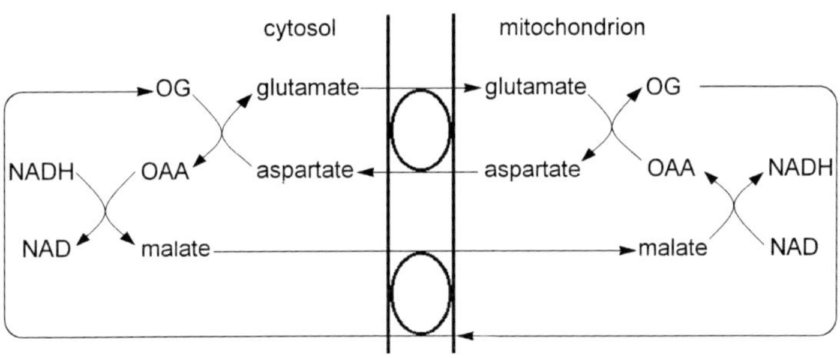

Figure 2. Scheme of the malate-aspartate shuttle. Abbreviations: OG, 2-oxoglutarate: OAA, oxaloacetate.

VI.3. Ex Vivo Effects of Ammonia on Mitochondrial Enzymes of Malate-Aspartate Shuttle

Hyperammonemia induces decreases in malate and succinate dehydrogenase activities in rat brain non-synaptic mitochondria. Activities of glutamate dehydrogenase and aspartate aminotransferasc in both mitochondria did not change in hyperammonemia [14, 54, 89]. Activities of all the enzymes above in the cytosol are unchanged in hyperammonemia. indicating that ammonia is the specific inhibitor of mitochondrial dehydrogenases and and explaining its inhibitory effects on mitochondrial respiration.

In synaptosomes and mitochondria isolated from brains of animals administered with acute dose of ammonium acetate, there is an increase in the activities of pyruvate, isocitrate, 2-oxoglutarate and succinate dehydrogenases while the changes in the activities of NAD-malate dehydrogenase, aspartate and alanine amino transferases were suppressed [89].

The activities of branched-chain amino acid transaminase and branched-chain keto acid dehydrogenase in mitochondria isolated from the rat cerebral cortex are not adversely affected in acute hyperammonemia [4].

VI.4. Ex Vivo Effects of Ammonia on Mitochondrial Metabolites of Malate-Aspartate Shuttle

The ammonia content of brain mitocbondria increased by 5-fold in rats injected with amnlonium acetate [54]. The anmlonium ion concentration in the mitochondrial water as calculated by an empirical formula [C] = 1.33 x C, where C is the mitochondrial content of the ammonium ion in nmol/mg of protein [37], was about 12 mM and 60 mM in control and hyperammonemic rats, respectively. The glutamate and aspartate contents decrease about 50% in brain mitochondria from hyperammonemic rats compared to corresponding controls; the malate and 2-oxoglutarate levels are similar in brain mitocllondrial preparations from control and hyperarnmonemic animals [54].

Collectively, Sections VI-2-4 indicate that the most probable controlling factor of the malate-aspartate shuttle in hyperammonemia seems to be the glutamate-aspartate exchange carrier .

VI.5. Effects of Ammonia on Mitochondrial Membrane Potential *In Vivo*

Energy state of mitochondria is determined by their ability to support the transmembrane potential difference, or mitochondrial membrane potential, $\Delta\psi$. Altered mitochondrial function is a crucial step in some mechanisms of cellular apoptosis. Accumulation of calcium in mitochondria may lead to the opening of the mitochondrial permeability transition pore (MPP), that contributes to both apoptosis and to necrotic cell death. Opening of MPP causes a dissipation of $\Delta\psi$.

It was found on astrocyte cultures using confocal microscopy and flow cytometry that $\Delta\psi$ decreased with ammonium chloride concentration, and this linear dependence was sensitive to cyclosporin A, a blocker of MPP. These results were interpreted as an ability of high ammonia levels to induce the MPP expression [6, 87]. However, *ex vivo* experiments did not supported this suggestion, ammonia injection to animals did not affect $\Delta\psi$ in isolated brain non-synaptic mitochondria as measured with tetraphenyl phosphonium cation [41, 63].

Disagreement between the absence the effect of ammonia on $\Delta\psi$ ex vivo [41, 63] and pronounced and dose-dependent inhibitory effect of ammonia on $\Delta\psi$ in vitro [6, 87] suggest that either common optical methods (confocal microscopy, flow cytometry) and potentiometry are irrelevant for the $\Delta\psi$ measurement, or the theoretical conception on the nature of MPP is false: MPP is a pore that can be or not be, can be only open or closed, but cannot be "self-opened" and, moreover, "dose-dependent". Alternatively, the regulation of MPP may be signified by the induction of additional pore expression and repression of active pores on the outer mitochondrial membrane, that can in turn be interpreted as opening or closing MPP. If so, ammonia is an effector of MPP expression, although no evidence is available in the literature.

Invariability of $\Delta\psi$ in ex *vivo* experiments [57] suggests that acute ammonia intoxication does not result in MPP expression in brain mitochondria. The results shows that non-synaptic brain mitochondria 1) do not open MPP under conditions favourable for opening MPP in rat liver mitochondria [8] and in synaptic brain mitochondria [11]; 2) are more resistant to MPP formation than rat liver and heart mitochondria; 3) preserve $\Delta\psi$ even after treatment with fast-acting lethal agents such as ammonia, which induce a damage to brain energy metabolism [54].

VI.6. Calcium Transport Across Rat Brain Mitochondrial Membrane

Calcium signalling system controls majority of cellular functions. Millimolar concentrations of calcium inhibit the key glycolytic and gluconeogenic enzymes in the cytosol [103], micromolar calcium activates several important dehydrogenases in the mitochondrial matrix [21, 22]. In neuronal cells, calcium signals govern a host of Ca^{2+}-dependent enzymes [23, 107].

Three main Ca^{2+} transport system are identified in mitochondria. The most active electrogenic Ca^{2+},transporter catalyzes Ca^{2+} uptake by mitochondria against the concentration gradient [91]. Выход Ca^{2+} efflux from mitochondria can involve $Ca^{2+}/2Na^{+}$ antiporter [18, 19, 90] or $Ca^{2+}/2H^{+}$ antiporter [2, 26]. Na-independent Ca^{2+} efflux can occur in mitochondria of all tissues and is usually coupled with MPP opening.

Ammonium ion is an effective physiological regulator of brain mitochondrial Ca^{2+} transport. Effects of hyperammonemia *in vivo* on Ca^{2+} transport in non-synaptic mitochondria from rat brain were investigated only in a few studies [57, 60].

VI.6.1. Endogenous Calcium in Mitochondria

Acute intoxication with ammonia induced a significant 62% increase in the endogenous calcium content of brain mitochondria [57, 60]. It can be a consequence of either increased Ca^{2+} uptake, decreased Ca^{2+} efflux, or Ca^{2+} release from other intramitochondrial stores.

VI.6.2. Calcium Uptake by Mitochondria

The rate of the energy-dependent Ca^{2+} uptake by brain mitochondria from rats injected with ammonia is much lower than that from control animals [57, 60].

VI.6.3. Mitochondrial Calcium Capacity and Calcium Efflux from Mitochondria

The maximal amount of Ca^{2+} taken up and steady retained by mitochondria is considered the Ca^{2+} capacity of mitochondria. The calcium capacity of mitochondria was significantly reduced in rats injected with ammonia.

When the amount of Ca^{2+} ions taken up by mitochondria in the energy-dependent way is in excess to their Ca^{2+} capacity, then accumulated calcium will exit the loaded mitochondria spontaneously. The spontaneous Ca^{2+} efflux rate was higher in mitochondria from rats injected with ammonia than in mitochondria from control rats.

It should be noted that the spontaneous Ca^{2+} efflux was.independent of cyclosporin A both in control mitochondria and in those from rats injected with ammonia, indicating that MPP does not play a role in calcium efflux (under the conditions studied). The spontaneous release of calcium from mitochondria is not affected by diltiazem or clonazepam, inhibitors of the Na^+/Ca^{2+} exchange, thus indicating that spontaneous release takes place by a mechanism which does not involve Na^+/Ca^{2+} exchange nor MPP [57, 60].

The dependence of the rate of Na^+-dependent Ca^{2+} efflux from rat brain non-synaptic mitochondria on Ca^{2+} content of these mitochondria is linear, however the slope of this dependence decrease as much as 2-fold in hyperammonemia [57, 60]. Thus, the activity of Na^+-dependent Ca^{2+} efflux from mitochondria decreases two-fold in hyperammonemia irrespective of intramitochondrial Ca^{2+} levels. The rate of spontaneous Na^+-independent Ca^{2+} efflux from rat brain non-synaptic mitochondria is increases about 2-fold in hyperammonemia [57, 60]. Both spontaneous Na^+-independent and Na^+-dependent Ca^{2+} effluxes in rat brain non-synaptic mitochondria are insensitive to cyclosporin A, a specific blockator of MPP, and hence take place using permeability chanells other than MPP. Chanells for Na^+-independent and Na^+-dependent Ca^{2+} effluxes differ one from another by sensitivity to ammonia, the first being activated 2-fold and the second being inactivated 2-fold [57, 60]. Alternatively, there can be presented a second MPP type totally insensitive to cyclosporin A. Such proposition was done in respect to rat liver mitochondria [33]. However, as it was said above, MPP can be only blocked, not self-closed with any effector and hence data available on mitochondrial Ca^{2+} transport contradict this hypothesis.

VI.6.4. t-Butyl Hydroperoxide-Induced Calcium Efflux from Mitochondria

A special, secondary Ca^{2+} transport system of *tert*-butyl hydroperoxide(tBH)-induced calcium efflux presents in the mitochondrion [94]. Agents affecting the redox state of mitochondrial pyridine nucleotides are known to cause changes in the rate and direction of Ca^{2+} movement across the inner mitochondrial membrane [67]. tBH induces redox changes in mitochondria by acting at the level of glutathione peroxidase. Addition of tBH to Ca^{2+}-loaded mitochondria results in 2-fold stimulation of Ca^{2+} efflux from control mitochondria but did not change the rate of Ca^{2+} efflux in mitochondria from hyperammonemic animals. This suggests that the impairment of the tBH-induced release of calcium in mitochondria from rats injected with ammonia may be due to the reduced activity of glutathione peroxidase. Externally added tBH seems to cannot be reduced adequately in mitochondria from rats injected with ammonia due to decreased glutathione peroxidase activity and GSH level. This may explain the inhibition of tBH-stimulated Ca^{2+} efflux in rats injected with ammonia. Cyclosporin A did not affect tBH-induced Ca^{2+} efflux and this process was not associated with swelling of mitochondria from control or hyperammonemic rats [57, 60].

VI.6.5. Comparison of Ammonia Effects In Vivo and In Vitro on Calcium Fluxes in Mitochondria

Effects of ammonia on Ca^{2+} uptake by brain non-synaptic mitochondria in vivo and in vitro are similar. In both models, ammonia induces a decrease in the maximum rate of Ca^{2+} uptake by approximately 40%. Effects of ammonium acetate and ammonium chloride are identical [57, 60], and hence ammonium ion is the main effector. Thus, ammonium ion *in vivo*, as well as *in vitro*, is the regulator of Ca^{2+} fluxes in rat brain mitochondria. Ammonia administered to animals is partly accumulated by brain mitochondria, decreases their ability to uptake Ca^{2+}, the rate of Na^{+}-dependent Ca^{2+} efflux from mitochondria and increases activity of the Na^{+}-independent, cyclosporin A-insensitive chanell of spontaneous Ca^{2+} efflux from brain non-synaptic mitochondria.

VII. Brain Lactate Accumulation Ex Vivo

Impairment of oxidative metabolism can be accopmpanied by inhibition of the NADH oxidation in the mitochondrial respiratory chhain and accumulation of reduced metabolitov such as lactate, malate, glutamate and isocitrate in the cell even even in the presence of sufficient oxygen supply [79]. The lactate/pyruvate ratio in rat brain increases while the glutamate/2-oxoglutarate ratio decreases in acute ammonia intoxication [78]. A hypothesis of accumulation of reduced metabolites in brain in acute hyperammonemia *in vivo* has been tested and confirmed in the only study [52]. Therein the lactate content of the brain increased up to 3-fold after the lethal dose of ammonium acetate.

The correlation between brain ammonia and lactate contents was linear under different conditions: in control, acute hyperammonemia, and after atropine and d-tubocurarine injection to hyperammonemic rats, with r=0.885 [53]. These data showed that ip injection of ammonia induced an intense increase in brain lactate, the brain lactate correlates with brain ammonia, and choline receptors are involved in ammonia and lactate accumulations.

VIII. Relationship of Acute Hyperammonemia and Oxidative Stress

A superabundance of free superoxide radicals (O_2^-) are dangerous to the cell. Any disturbance of the electron transfer in the mitochondrial respiratory chain leads to increased O_2^- generation [13]. The mitochondrion is also the main intracellular generator of H_2O_2 [10]. H_2O_2 ois formed in the chemical reaction of O_2^- dismutation calalyzed by Mn^{2+}-dependent superoxide dismutase (Mn-SOD) in the mitochondrial matrix as well as in monoamine oxidase (MAO) reaction on the outer mitochondrial membrane. The O_2^- and H_2O_2 contents of mitochondria and hence the toxic effects of these reactive oxygen species (ROS) is depend, directly or indirectly, on the activities of the respiratory chain, H_2O_2-producing enzymes and H_2O_2-consuming catalase and glutathione peroxidase.

VIII.1. Ex Vivo Effects of Ammonia on Superoxide and Hydrogen Peroxide Production

The rate of O_2^- production by submitochondrial particles from brains of acute hyperammonemic rats was increased 100% as compared with control levels [55, 60]. The rate of H_2O_2 production by intact mitochondria was decreased under similar conditions. The mitochondrial respiratory chain is commonly believed to generate H_2O_2 [106]. However, studies performed on the hyperammonemic animal model have shown that H_2O_2 is produced by intact rat brain non-synaptic mitochondria but did not by submitochondrial particles prepared from the same mitochondria [60].

As both preparations contain the respiratory chain and only intact mitochondria contain the matrix Mn-SOD activity, these results suggest that H_2O_2 cannot be formed by the respiratory chain but is formed by the Mn-SOD activity.

In conclusion, the Mn-SOD activity is the only sourse of H_2O_2 in rat brain non-synaptic mitochondria. So, the rate of H_2O_2 production by mitochondria decreases ib acute hyperammonemia in parallel with a decrease in Mn-SOD activity [60].

VIII.2. Effects of Ammonia on Pro- and Antioxidant Enzyme Activities

The only scientific group studied the in vivo effects of acute ammonia intoxication on prooxidant and antioxidant enzyme activities in brain mitochondria [48, 49, 55, 56, 60, 62; 104). These workers found that the MAO activity in isolated rat brain non-synaptic mitochondria increased and activities of Mn-SOD, catalase and glutathione peroxidase decreased in acute hyperammonemia.

VIII.3. A Summary of Ex Vivo Effects of Ammonia on Brain Mitochondria

Summary of effects of acute hyperammonemia in rats on character parameters of oxidative stress in brain mitochondria is given in the table. All parameters change catastrofically, approximately 2-25-fold in hyperammonemia and strongly towards the increased oxidative stress [61].

In spite of these disorders, functional properties of mitochondria persist. Non-synaptic rat brain mitochondria are more resistant to MPP formation, Ca^{2+}-induced swelling, and dissipation of $\Delta\psi$ under conditions of ammonia-stimulated ROS overproduction, excess Ca^{2+} accumulation, increased oxidative stress and disturbed energy metabolism, than liver and heart mitochondria [41].

Table. Effects of ammonium acetate injection on some parameters of oxidative stress in rat brain mitochondria (as calculated from data of [60, 61])

Parameter	Ammonia/control, %
O_2^- production	191
H_2O_2 production	62
GSH/GSSG ratio	53
Free NAD/NADH ratio	2370
Free NADP/NADPH ratio	2557
Mn-SOD activity	68
Catalase activity	53
Glutathione peroxidase activity	68
Monoamine oxidase activity	159

Thus, $\Delta\psi$ does not change in rat brain mitochondria after acute injection of a lethal dose of ammonia. These data show that acute ammonia intoxication in rats in vivo did not lead to the formation of the MPP and mitochondrial swelling despite a significant increase in the content of Ca^{2+} in brain mitochondria [57].

The nonsynaptic mitochondrial preparation is largely composed of astrocytic mitochondria. It has been reported that exposure of cultured astrocytes to large concentrations of ammonia induced changes in $\Delta\psi$ and MPP [74]. These changes do not occur in the brain mitochondria ex vivo [63]. The in vitro exposure of astrocytes to ammonia lasted for a long time and the effects observed were likely mediated by accumulating glutamine [3]. The lack of effect ex vivo was most likely due to the short time of exposure.

IX. Markers of Neuronal Apoptosis

Oxidative stress in mitochondria represents one of initial steps of cell death. Mitochondria can itself integrate various deadly events. Accumulation of Ca^{2+} in mitochondria may activate a complex molecular mechanism united into a conception of MPP, that contributes to both apoptosis and to necrotic cell death [44, 68]. Increased ROS may lead to formation of MPP, dissipation of $\Delta\psi$, an increase in the matrix volume, mitochondrial swelling, mechanical disruption of the outer mitochondrial membrane and release of mitochondrial factors that induce apoptosis [68]. including cytochrome c.

IX.1. Efflux of Cytochrome *C* into Cytoplasm

Mitochondria play a critical role in the cell death pathway by releasing signaling proteins from their intermembrane space into the cytosol [46]. It has been postulated that in response to an suicidal stimulus mitochondria releases cytochrome c into the cytoplasm where it interacted with Apaf-1 in the apoptosome complex [69. 70, 80] leading to apoptosis. It has been postulated that the leakage of cytochrome c from mitochondria resulted from the opening of MPP [82, 95]. Cytochrome c can be released from brain mitochondria by an MPP-independent mechanism and without the $\Delta\psi$ alteration [46, 83].

No evidence for a role of cytochrome c release in acute ammonia toxicity was found. A significant decrease (30%) of cytochrome c content in brain mitochondria from rats injected with ammonia was observed without increase in cytochrome c in the cytosol [63]. Analogous results were obtained in mitochondria from Jurkat cells, undergoing Fas-mediated apoptosis [64] At present, mechanisms responsible for this phenomenon are not clear. One of the reason for such the pattern can be ammonia-induced activation of mitochondrial proteases and corresponding cytochrome c cleavage.

Cytochrome c is located in the mitochondrial intermembrane space and is the essential component of the mitochondrial respiratory chain. The decrease in cytochrome c in mitochondria may contribute to the reduced state 3 respiration, decreased respiratory control index and disturbances in the mitochondrial electron transport chain reported previously in brain mitochondria from rats injected with ammonia [54].

IX.2. Caspases in Mitochondria, Cytoplasm and Nuclei

Released cytochrome c can trigger the proteolytic maturation of caspases within the apoptosome, that includes Apaf-1, caspase-9, and caspase-3 and cytochrome c [69, 70, 80], or can activate caspase-independent cell death pathways through apoptosis initiating factor, AIF [100].

Mitochondrial dysfunction may result in induction of cytosolic cysteine proteases caspase 9 and caspase 3. Both enzymes are responsible for the later steps of apoptosis [5]. After binding with apoptosis protease activating factor-1 (Apaf-1) which requires the presence of cytochrome c and ATP (or dATP) in the cytosol (i.e., after apoptosome formation), pro-caspase 9 is activated to caspase 9, and the latter becomes capable of cleaving and activating caspase 3. Caspase 3 can participate in DNA fragmentation directly [36] or activate endonucleases such as caspase-activated DNase cleaving chromatin and, as a consequence, execute apoptosis.

The effect of acute hyperammonemia on the two caspases was studied recently [41, 63]. Acute ammonia intoxication did not affect caspase-9 or caspase-3 activities.

IX.3. Apoptotic Alterations in Cell Nuclei

IX.3.1. DNA Fragmentation

Not only release of cytochome c and caspase 3 activation but also nuclear DNA fragmentation are markers of apoptosis. Acute ammonia intoxication leads to the 44-fold increase in ammonia levels in nuclei of brain cells and to early formation of internucleosomal DNA damage in nuclei [62]. This suggests that ammonia could activate apoptotic pathways involving altered mitochondrial function and DNA damage. Apoptotic death is not, however, usually found in hyperammonemic states [63].

IX.3.2. Poly(ADP-Ribose) Polymerase Levels and Changes

Poly(ADP-ribose) polymerase (PARP) activity was considered as another marker of apoptosis. Poly(ADP-ribose) polymerase (PARP) is a multifunctional enzyme located in the nucleus of cells in various organs including the brain [35]. This enzyme is involved in specific cellular functions such as DNA repair [93]. Intact mammalian PARP of Mr 116,000 is ptoteolytically cleaved by caspases to a fragment of 85,000 during apoptosis [66] or additionally fragments of 35,000-40,000 and 50,000 during necrosis [98]. Acute ammonia intoxication leads to a significant increase in PARP content in nuclei of brain cells by 100% that would be associated with increased PARP activity [62]. Ammonia-induced increase in PARP is dependent on de novo protein synthesis as indicated by the prevention of this increase by cycloheximide. Immunoblotting performed with a PARP monoclonal antibody did not detect bands corresponding to any lowmolecular fragment [62].

Hence, PARP is synthesized de novo and do not cleaved following acute ammonia injection.

IX.3.3. Nuclear NAD Levels

PARP is activated in response to DNA damage and synthesizes and transfers negatively charged polymers of ADP-ribose to chromatin-associated proteins, using NAD as a substrate. When DNA damage is extensive, excessive poly(ADP-ribose) formation by PARP may deplete cellular NAD pools [34, 97]. NAD synthetase is then activated to synthesize NAD in an energy-dependent manner and can result in ATP depletion. Impairment of intracellular energy metabolism by excessive PARP activation may contribute to cell death [7].

Ammonia injection to animals leads to a significant increase (approximately 200% of control) in the content of PARP in nuclei of brain cells [62].

IX.3.4. Nuclear NAD Synthetase and NAD Glycohydrolase Activities

Acute ammonia intoxication alters nuclear NAD synthetase activity but not NAD glycohydrolase activity. The decrease in NAD content in nucleus may be due to a decrease in its synthesis by NAD synthetase or to an increase in its degradation by NAD glycohydrolase or in its consumption in other reactions. To help to clarify this point we measured the activity of both enzymes in nucleus of brain cells from rats injected or not with ammonia. extremely low basal activity of NAD synthetase in brain nuclei of control rats increased significantly 5 min after injection of ammonia. The increase in NAD synthetase activity was transient and the activity of the enzyme decreased at 8 min and was not detectable at 11 min after ammonia injection [62]. NAD hydrolase activity in nuclei was not affected by injection of ammonia [62].

IX.3.5. P53 Dynamics

One possible explanation for the absence of apoptosis could be altered localization of p53. Cytosolic localization of p53 seems necessary and sufficient to induce apoptosis [15]. Qu et al. [85] also found that endoplasmic reticulum stress inhibited p53-mediated apoptosis and that increased cytoplasmic localization of an inactive phosphorylated form of p53 was involved in the mechanism of the inhibition. Caspase-independent death mechanisms are modulated by the presence p53, the tumor suppressor protein, which acts as a key regulator of neuronal death after acute injury such as DNA damage [42]. p53 protein is believed to be a potentially important downstream target of ammonia neurotoxicity [81] and PARP involved in the regulation of p53 [108]. P53 is cytosolic enzyme in the most cell types [92] and exists in a latennt, inactive form [65]. Some stress situations may induce the formation of active p53 [84], which then is transferred from the cytosol into the nucleus. It was found, using ELISA that ammonia injection to rats results in a significant increase in brain cytosolic p53 levels, by 100–120% of controls. However, the nuclear and mitochondrial p53 levels were unchanged in acute ammonia intoxication [41, 63] when cytosolic p53 was increased and early nuclear DNA damage was observed [62]. Chipuk et al. [15] proposed that cytosolic localization of p53 seems necessary and sufficient to induce apoptosis. Qu et al. [85] found that endoplasmic reticulum stress inhibited p53-mediated apoptosis and that increased cytoplasmic localization of an inactive phosphorylated form of p53 was involved in the mechanism of the inhibition.

Collectively, these data confirm a new theory according to which inactivation of p53 in nuclei completely protect these cells from apoptosis [20]. There is therefore no evidence for a role of apoptosis in acute ammonia toxicity. The resistance of nuclei to ammonia-induced apoptosis could be due to disturbed p53 translocation from the cytosol into nuclei.

X. Biochemical Processes in the Cytoplasm

X.1. Cytosolic Pro-Oxidant Enzymes

Among cytosolic ROS-generating enzymes, xanthine oxidase, aldehyde oxidase and Cu,Zn- SOD play important roles. Monoamine oxidase located on the outer mitochondrial membrane and releases ROS into the cytosol, too.

Xanthine:dehydrogenase (XD) and xanthine oxidase (XO), two enzyme forms of the XD/XO enzyme complex, are the end steps in the purine catabolic pathway and directly involved in depletion of the adenylate pool in the cell. XD and XO have also been implicated as a source of reactive oxygen species (ROS) inducing neuronal cell injury [30].

XD predominates in healthy tissue, but under pathological conditions XD may be readily converted to XO through the reversible thiol oxidation of sulfhydryl residues on XD or by the irreversible proteolytic cleavage of a fragment of XD [17, 77].

Injection of rats with ammonium acetate caused an inhibition of XD activity in the brain cytosol and increase in XO activity and the XO/XD activity ratio suggesting the conversion of XD to XO and the increase in ROS production [39, 59].

Increase in the activities of cytosolic aldehyde dehydrogenase and mitochondrial monoamine oxidase following acute injection of ammonium acetate [59, 61] suggests elevated ROS production. Intracellular Ca^{2+} do activate phospholipase A_2 in the brain [99], as well as oxidative arachidonate metabolism [1], both accompanied with excessive ROS production.

Ammonia injection leads to increase in ADP and AMP levels and to a decrease in ATP and the total adenylate pool. Brain xanthine and hypoxanthine levels increased 2-5-fold in acute hyperammonemia [39]. Activities of AMP deaminase and adenosine deaminase, the key enzymes of adenine nucleotide breakdown pathway, increase significantly in brain regions [40].

Thus, acute ammonia intoxication favours to accelerated breakdown of adenine nucleotides and increased oxidative stress in the neuronal cell cytoplasm.

X.2. Role of Nitric Oxide in Changes of Activity of Antioxidant Enzymes

The mechanisms by which overactivation of NMDA receptors leads to neuronal degeneration and death involve activation of NO synthase [45] and the formation of nitric oxide (NO) [71].

Inhibitors of NO synthase such as nitroarginine prevent ammonia toxicity and ammonia-induced alterations in brain energy metabolism [52] and in activities liver and brain antioxidant enzymes [56].

Thus, the NO/NO synthase system is involved in the mechanism of ammonia toxicity.

XI. Alterations in the Plasma Membrane

XI.1. Brain ATPase in Acute Ammonia Intoxication

The neuronal plasma membrane contains Na,K-ATPase, glutamate receptors and other proteinaceous structures involved in energy metabolism. Na,K-ATPase supports sodium and potassium ion balance across the membrane using chemical energy of ATP hydrolysis while the glutamate receptor of NMDA type is an essential component of the synaptic glutamatergic neurotransmission. Activation of NMDA receptors is coupled with opening own ionic chanell allowing to penetrate Ca^{2+} and Na^{+} into the postsynaptic neuron.

Brain Na,K-ATPase activity increased in acute ammonia intoxication and this effect depends on the function of NMDA receptors [51].

XI.2. Roles of NMDA Receptors in Acute Ammonia Toxicity

The role of NMDA receptors in ammonia toxicity is widely described (e.g. [73]). Usually, It is usually studied using (+)-5-methyl-10,11-dihydro-5H-dibenzo[a,d]cyclopenten-5,10-imine hydrogen maleate (MK-801), a specific antagonist of these receptors. MK-801 injection to animals before lethal dose of ammonia increase survival [51]. Blocking NMDA receptors impedes ammonia-induced alterations in energy metabolism and antioxidant defence system. MK-801 application completely prevents changes of intracellular concentrations of ammonia, lactate, pyruvate, acetoacetate [51], depletion of ATP, accumulation of ADP, AMP, xanthine and hypoxanthine, a decrease in adenylate pool size in brain tissue, acceleration of O_2^- production, decrease in Mn-SOD, catalase and glutathione peroxidase activities and increase in monoamine oxidase A activity in nonsynaptic brain mitochondria, a XD to XO conversion, increase in XO and aldehyde dehydrogenase activities in brain cytosol [39, 59]. The ammonia-induced increase in intramitochondrial calcium and spontaneous release of calcium from mitochondria [57] and depletion of nuclear NAD [62] are completely prevented by previous blocking of NMDA receptors with MK-801. Tissue concentrations of glycogen, glucose, 3-hydroxybutyrate, glutamate, glutamine and inorganic phosphate in the brain are partly repaired after theatment of hyperammonemic animals with MK-801 [51].

Collectively, data above suggest that NMDA receptors are involved in alterations of energy metabolism and antioxidant status of the brain underlying acute ammonia toxicity.

CONCLUSION

The data reviewed here provide substantial evidence that ammonia impacts multiple biochemical processes in the animal body. Its effects are an inhibition of hepatic gluconeogenesis and ketogenesis and brain aerobic glucose oxidation (Figure 3). Ammonia is an inhibitor of the mitochondrial respiratory chain, malate-aspartate shuttle and a complex effector of calcium transport. Ammonia induces overactivation of glutamate NMDA receptors on the postsynaptic plasma membrane allowing calcium and sodium cations to enter the neuron. Increased cellular calcium concentration activates calcium-dependent enzymes such as phospholipase A_2 and NO synthase. Oxidative metabolism of arachidonate generates ROS while NO synthase produces NO radical, so depressing activities of all antioxidant enzymes. The subsequent increase in membrane lipid peroxidation, damage to the plasma membrane and cell death occur.

Toxic effects of ammonia is spreading all over neuronal intracellular compartments, including plasma membrane, mitochondria, cytosol, and nucleus. The results reported are summarized in Figure 4. Ammonia causes alterations of concentrations of glycolytic intermediates and end products, adenine nucleotides and amino acidshe brain tissue, a decrease in activities of antioxidant enzymes such as Mn-SOD, catalase and glutathione peroxidase in brain mitochondria, an increase in activities of pro-oxidant monoamine oxidase in mitochondria, xanthine and aldehyde oxidases in the brain cytosol, as well as stimulates ROS formation in mitochondria, cytosol and nuclei.

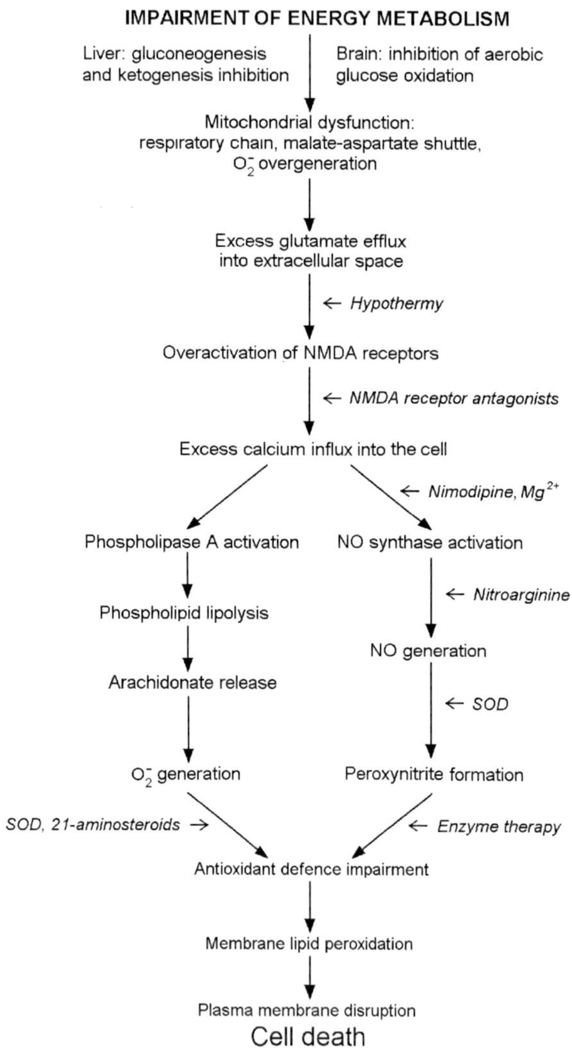

Figure 3. Scheme of proposed events in acute hyperammonemia: from impairment of energy metabolism to cell death. Arrows ← indicate points of the action and effectors impeding the process.

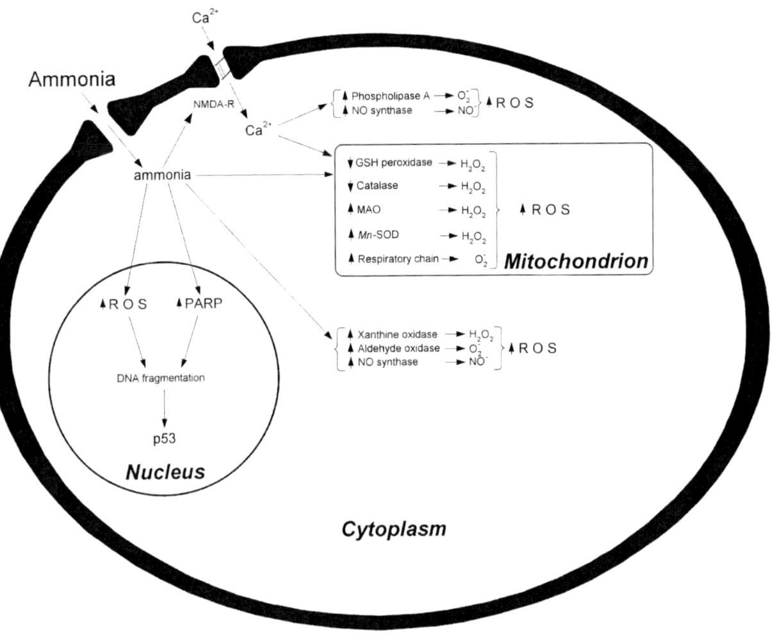

Figure 4. Scheme of biochemical alterations in a neuronal cell in acute ammonia intoxication.

Acute ammonia intoxication leads to nuclear DNK fragmentation and to activation of a nuclear DNA repair enzyme PARP. Ammonia-induced changes in energy and oxidative metabolism completely or partly prevented by MK-801 or nitroarginine, suggesting that NMDA receptors and NO synthase are involved in mechanisms of acute ammonia toxicity. Ammonia toxicity in vivo is not related with caspases 9 and 3 induction, dissipation of mitochondrial membrane potential, cytochrome c release from non-synaptic brain mitochondria, and induction of apoptotic markers in nuclei. There is no evidence for the involvement of mitochondria in neuronal apoptosis in acute hyperammonemia.

References

[1] Adibhatla, RM; Hatcher, JF. Phospholipase A(2), reactive oxygen species, and lipid peroxidation in CNS pathologies. *BMB Rep.*, 2008, 41, 560–567.

[2] Akerman, KE. Effect of pH and Ca^{2+} on the retention of Ca^{2+} by rat liver mitochondria. *Arch. Biochem. Biophys*. 1978, 189, 256–262.

[3] Albrecht, J; Norenberg, MD. Glutamine: a Trojan horse in ammonia neurotoxicity. *Hepatology*, 2006, 44, 788–794.

[4] Alexander, J; Murthy, CR. Alterations in mitochondrial branched-chain amino acid metabolism in brain in acute hyperammonemic states. *Neurosci. Lett.*, 1993, 149, 221–224.

[5] Alnemri, ES; Livingston, DJ; Nicholson, DW; Salvesen, G; Thornberry, NA; Wong, WW; Yuan, J. Human ICE/CED-3 protease nomenclature. *Cell*, 1996, 87, 171.

[6] Bai, G; Rama Rao, KV; Murthy, CR; Panickar, KS; Jayakumar, AR; Norenberg, MD. Ammonia induces the mitochondrial permeability transition in primary cultures of rat astrocytes. *J. Neurosci. Res.*, 2001, 66, 981–991.

[7] Berger, N.A. Poly (ADP-ribose) in the cellular response to DNA damage. *Radiat. Res*., 1985, 101, 4-15.

[8] Berman, SB; Watkins, SC; Hastings, TG. Quantitative biochemical and ultrastructural comparison of mitochondrial permeability transition in isolated brain and liver mitochondria: evidence for reduced sensitivity of brain mitochondria. *Exp Neurol*, 2000, 164, 415–425.

[9] Bessman, SP; Bessman, AN. The cerebral and peripheral uptake of ammonia in liver diseasewith an hypothesis for the mechanism of hepatic coma. *J Clin Invest*, 1955, 34, 622–628.

[10] Boveris, A. Mitochondrial production of superoxide radical and hydrogen peroxide. *Adv. Exp. Med. Biol.*, 1977, 78, 67–82.

[11] Brown, MR; Sullivan, PG; Geddes, JW. Synaptic mitochondria are more susceptible to Ca^{2+} overload than nonsynaptic mitochondria. *J. Biol. Chem.*, 2006, 281, 11658–11668.

[12] Cande, C; Cecconi, F; Dessen, P; Kroemer, GJ. *Cell Sci.*, 2002, 115, 4727–4734.

[13] Cheah, KS; Chance, B. The oxidase systems of *Ascaris*-muscle mitochondria. *Biochim. Biophys. Acta*, 1970, 223, 55–60.

[14] Cheeseman, AJ; Clark, JB. Influence of the malate-aspartate shuttle on oxidative metabolism in synaptosomes. *J. Neurochem.*, 1988, 50, 1559–1565.

[15] Chipuk, JE; Kuwana, T; Bouchier-Hayes, L; Droin, NM; Newmeyer, DD; Schuler, M; Green, DR. Direct activation of Bax by p53 mediates mitochondrial membrane permeabilization and apoptosis. *Science*, 2004, 303(5660), 1010–1014.

[16] Cooper, AJ; Plum, F. Biochemistry and physiology of brain ammonia. *Physiol. Rev.*, 1987, 67, 440–519.
[17] Corte, ED; Stirpe, F. The regulation of rat liver xanthine oxidase. Involvement of thiol groups in the conversion of the enzyme activity from dehydrogenase (type D) into oxidase (type O) and purification of the enzyme. *Biochem. J.*, 1972, 126, 739–745.
[18] Crompton, M; Heid, I. The cycling of calcium, sodium, and protons across the inner membrane of cardiac mitochondria. *Eur. J. Biochem.*, 1978, 91, 599–608.
[19] Crompton, M; Moser, R; Ludi, H; Carafoli, E. The interrelations between the transport of sodium and calcium in mitochondria of various mammalian tissues. *Eur. J. Biochem.*, 1978, 82, 25–31.
[20] Cui, H; Schroering, A; Ding, HF. p53 mediates DNA damaging drug-induced apoptosis through a caspase-9-dependent pathway in SH-SY5Y neuroblastoma cells. *Mol. Cancer Ther.* 2002, 1, 679–686.
[21] Denton, RM; McCormack, JG. On the role of the calcium transport cycle in heart and other mammalian mitochondria. *FEBS Lett.*, 1980, 119, 1–8.
[22] Denton, RM; McCormack, JG. Ca^{2+} transport by mammalian mitochondria and its role in hormone action. *Am. J. Physiol.*, 1985, 249,.E543–E554.
[23] Farooqui, AA; Horrocks, LA. Excitatory amino acid receptors, neural membrane phospholipid metabolism and neurological disorders. *Brain Res. Rev.*, 1991, 16, 171–191.
[24] Felipo, V. Hyperammonemia. *Handbook of Neurochemistry and Molecular Neurobiology*, 2009, 43–69.
[25] Felipo, V; Butterworth, RF. Mitochondrial dysfunction in acute hyperammonemia. *Neurochem. Int.*, 2002, 40, 487–491.
[26] Fiskum, G; Cockrell, RS. Ruthenium red sensitive and insensitive calcium transport in rat liver and Ehrlich ascites tumor cell mitochondria. *FEBS Lett.*, 1978, 92, 125–128.
[27] Fitzpatrick, SM; Cooper, AJL; Duffy, TE. Use of β-methylene-D,L-aspartate to assess the role of aspartate aminotransferase in cerebral oxidative metabolism. *J. Neurochem.*, 1983, 41, 1370–1383.
[28] Gimpel, JA; Haan, EJ; Tager, JM. Permeability of isolated mitochondria to oxaloacetate. *Biochim. Biophys. Acta*, 1973, 292, 582–591.
[29] Hahn, M; Massen, O; Nencki, M; Pavlov, I. Die Eck-sche Fistel zwischen der unteren Hohlvene und der Pfortader und ihre Folgen fur den Organismus. *Arch. Exp. Path Pharmakol.*, 1893, 32, 161–210.

[30] Haorah, J; Ramirez, SH; Floreani, N; Gorantla, S; Morsey, B; Persidsky, Y. Mechanism of alcohol-induced oxidative stress and neuronal injury. *Free Radic. Biol. Med.*, 2008, 45, 1542–1550.

[31] Häussinger, D; Görg, B. Interaction of oxidative stress, astrocyte swelling and cerebral ammonia toxicity. *Curr. Opin. Clin. Nutr. Metab. Care*, 2010, 13, 87–92.

[32] Hawkins, RA; Miller, AL; Nielsen, RC; Veech, RL. The acute action of ammonia on rat brain metabolism in vivo. *Biochem. J.*, 1973, 134, 1001–1008.

[33] He, L; Lemasters, JJ. Regulated and unregulated mitochondrial permeability transition pores: a new paradigm of pore structure and function? *FEBS Lett.*, 2002, 512, 1–7.

[34] Heller, B; Wang, ZQ; Wagner, EF; Radons, J; Burkle, A; Fehsel, K; Burkart, V; Kolb, H. Inactivation of the poly(ADP-ribose) polymerase gene affects oxygen radical and nitric oxide. *J. Biol. Chem.*, 1995, 270, 11176–11180.

[35] Ikai, K; Ueda, K; Hayaishi, O. Immunohistochemical demonstration of poly(adenosine diphosphate-ribose) in nuclei of various rat tissues. *J. Histochem. Cytochem.*, 1980, 28, 670–676.

[36] Janicke, RU; Sprengart, ML; Wati, MR; Porter, AG. Caspase-3 is required for DNA fragmentation and morphological changes associated with apoptosis. *J. Biol. Chem.*, 1998, 273, 9357–9360.

[37] Kaminsky, YG; Kosenko, EA. Calculation of the concentration of metabolites freely and nonfreely penetrating the cell membranes in the liver cytosol and mitochondria. *Izv USSR Academy of Sciences, Ser. Biol.*, 1987, No 2, 196–202. In Russian.

[38] Kaminsky, YG; Kosenko, EA. *Paradoxes of carbohydrate metabolism.* Pushchino: ONTI; 1988. In Russian.

[39] Kaminsky, Y; Kosenko, E. Brain purine metabolism and xanthine dehydrogenase/oxidase activity in hyperammonemia are under control of NMDA receptors and nitric oxide. *Brain Res.*, 2009, 1294, 193–201

[40] Kaminsky, Y; Kosenko, E. AMP deaminase and adenosine deaminase activities in liver and brain regions in acute ammonia intoxication and subacute toxic hepatitis. *Brain Res.*, 2010, 1311, 175–181.

[41] Kaminsky, YG; Kosenko, EA; Venediktova, NI; Felipo, V; Montoliu, C. Apoptotic markers in the mitochondria, cytosol, and nuclei of brain cells during ammonia toxicity. *Neurochem. J.*, 2007, 1, 78–85.

[42] Kastan, MB; Onyekwere, O; Sidransky, D; Vogelstein, B; Craig, RW. Participation of p53 protein in the cellular response to DNA damage. *Cancer Res.*, 1991, 51, 6304–8311.

[43] Katunuma, N; Okada, M. Respiratory inhibition of TCA cycle and control of glutamic acid synthesis by ammonia in rat liver mitochondria. *Biochem. Biophys. Res. Commun.*, 1963, 12, 252–256.

[44] Kim, JS; He, L; Qian, T; Lemasters, JJ. Role of the mitochondrial permeability transition in apoptotic and necrotic death after ischemia/reperfusion injury to hepatocytes. *Curr. Mol. Med.*, 2003, 3, 527–535;

[45] Kitano, T; Matsumura, S; Seki, T; Hikida, T; Sakimura, K; Nagano, T; Mishina, M; Nakanishi, S; Ito, S. Characterization of N-methyl-D-aspartate receptor subunits involved in acute ammonia toxicity. *Neurochem. Int.*, 2004, 44, 83–90.

[46] Kluck, RM; Bossy-Wetzel, E; Green, DR; Newmeyer, DD. The release of cytochrome c from mitochondria: a primary site for Bcl-2 regulation of apoptosis. *Science*, 1997, 275, 1132–1136.

[47] Kosenko, EA; Kaminsky, YG. *Energy metabolism in rat tissues in acute ammonia intoxication*. Pushchino: ONTI; 1989. In Russian.

[48] Kosenko, EA; Kaminsky, YG. *Cellular mechanisms of ammonia toxicity*. Moscow:LKI; 2008. In Russian.

[49] Kosenko, EA; Kaminsky, YG. Brain monoamine oxidase A in hyperammonemia is regulated by NMDA receptors. *Centr. Eur. J. Biol.*, 2009, 4, 321–326.

[50] Kosenko, EA; Kaminsky, YG; Felipo, V; Minana, MD; Grisolia, S. Chronic hyperammonemia prevents changes in brain energy and ammonia metabolites induced by acute ammonium intoxication. *Biochim. biophys. Acta*, 1993, 1180, 321–326.

[51] Kosenko, E; Kaminsky, Y; Grau, E; Minana, MD; Marcaida, G; Grisolia, S; Felipo, V. Brain ATP depletion induced by acute ammonia intoxication in rats is mediated by activation of the NMDA receptor and Na^+,K^+-ATPase. *J. Neurochem.*, 1994, 63, 2172–2178.

[52] Kosenko, E; Kaminsky, Y; Grau, E; Minana, MD; Grisolia, S; Felipo, V. Nitroarginine, an inhibitor of nitric oxide synthetase, attenuates ammonia toxicity and ammonia-induced alterations in brain metabolism. *Neurochem. Res.*, 1995, 20, 451–456.

[53] Kosenko, EA; Kaminskii, YG; Korneev, VN; Lukyanova, LD. Protective action of M- and N-cholinoceptor blockers in acute ammonium intoxication. *Bull Exp. Biol. Med.*, 1995, 120, 1111–1114.

[54] Kosenko, E; Felipo, V; Montoliu, C; Grisolia, S; Kaminsky, Yu. Effects of acute hyperammonemia in vivo on oxidative metabolism in nonsynaptic rat brain mitochondria. *Metab. Brain Dis.*, 1997, 12, 69–82.

[55] Kosenko, E; Kaminsky, Y; Kaminsky, A; Valencia, M; Lee, L; Hermenegildo, C; Felipo, V. Superoxide production and antioxidant enzymes in ammonia intoxication in rats. *Free Radic. Res.*, 1997, 27, 637–644.

[56] Kosenko, E; Kaminsky, Y; Lopata, O; Muravyov, N; Kaminsky, A; Hermenegildo, C; Felipo, V. Nitroarginine, an inhibitor of nitric oxide synthase, prevents changes in superoxide radical and antioxidant enzymes induced by ammonia intoxication. *Metab. Brain Dis.*, 1998, 13, 29–41.

[57] Kosenko, E; Kaminsky, Y; Stavroskaya, IG; Felipo, V. Alteration of mitochondrial calcium homeostasis by ammonia-induced activation of NMDA receptors in rat brain in vivo. *Brain Res.*, 2000, 880, 139–146.

[58] Kosenko, E; Venediktova, N; Kaminsky, Y; Montoliu, C; Felipo, V. Preparation and handling of brain mitochondria useful to study uptake and release of calcium. *Brain Res. Brain Res. Protoc.*, 2001, 7, 248–254.

[59] Kosenko, E; Venediktova, N; Kaminsky, Y; Montoliu, C; Felipo, V. Sources of oxygen radicals in brain in acute ammonia intoxication in vivo. *Brain Res.*, 2003, 981, 193–200.

[60] Kosenko, EA; Venediktova, NI; Kaminsky, YG; Montoliu, C; Felipo, V; Косенко, EA. Calcium transport across the inner mitochondrial membrane in rat brain in hyperammonemia. *Biol. Membr. (Moscow)*, 2003, 20, 149–159. In Russian.

[61] Kosenko, EA; Venediktova, NI; Kaminsky, YG. Calcium and ammonia stimulate monoamine oxidase A activity in brain mitochondria. *Biol. Bull*, 2003, 30, 449-453.

[62] Kosenko, E; Montoliu, C; Giordano, G; Kaminsky, Y; Venediktova, N; Buryanov, Y; Felipo, V. Acute ammonia intoxication induces an NMDA receptor-mediated increase in poly(ADP-ribose) polymerase level and NAD metabolism in nuclei of rat brain cells. *J. Neurochem.*, 2004, 89, 1101–1110.

[63] Kosenko, E; Kaminsky, Y; Solomadin, I; Marov, N; Venediktova, N; Montoliu, C; Felipo, V. Acute ammonia neurotoxicity in vivo involves increase in cytoplasmic protein p53 without alterations in other markers of apoptosis. *J. Neurosci. Res.*, 2007, 85, 2491–2499.

[64] Krippner, A; Matsuno-Yagi, A; Gottlieb, RA; Babior, BM. Loss of function of cytochrome c in Jurkat cells undergoing fas-mediated apoptosis. *J. Biol. Chem.*, 1996, 271, 21629–21636.

[65] Lane, D. Awakening angels. *Nature*, 1998, 394, 616–617.

[66] Lazebnik, YA; Kaufmann, SH; Desnoyers, S; Poirier, GG; Earnshaw, WC. Cleavage of poly(ADP-ribose) polymerase by a proteinase with properties like ICE. *Nature*, 1994, 371, 346–347.

[67] Lehninger, AL; Vercesi, A; Bababunmi, EA. Regulation of Ca^{2+} release from mitochondria by the oxidation-reduction state of pyridine nucleotides. *Proc. Nat. Acad. Sci. USA*, 1978, 75, 1690–1694.

[68] Lemasters, JJ; Qian, T; Trost, LC; Herman, B; Cascio, WE; Bradham, CA; Brenner, DA; Nieminen, AL. Confocal microscopy of the mitochondrial permeability transition in necrotic and apoptotic cell death. *Biochem. Soc. Symp.*, 1999, 66, 205–222.

[69] Li, P; Nijhawan, D; Budihardjo, I; Srinivasula, SM; Ahmad, M; Alnemri, ES; Wang, X. Cytochrome c and dATP-dependent formation of Apaf-1/caspase-9 complex initiates an apoptotic protease cascade. *Cell*, 1997, 91, 479–489.

[70] Liu, X; Kim, CN; Yang, J; Jemmerson, R; Wang, X. Induction of apoptotic program in cell-free extracts: requirement for dATP and cytochrome. *Cell*, 1996, 86, 147–157.

[71] Mattson, MP; Lovell, MA; Furukawa, K; Markesbery, WR. Neurotrophic factors attenuate glutamate-induced accumulation of peroxides, elevation of intracellular Ca^{2+} concentration, and neurotoxicity and increase antioxidant enzyme activities in hippocampal neurons. *J. Neurochem.*, 1995, 65, 1740–1751.

[72] McKhann, GM; Tower, DB. Ammonia toxicity and cerebral energy metabolism. *Am. J. Physiol.*, 1961, 200, 420–424.

[73] Monfort, P; Kosenko, E; Erceg, S; Canales, JJ; Felipo, V. Molecular mechanism of acute ammonia toxicity: role of NMDA receptors. *Neurochem. Int.*, 2002, 41, 95–102.

[74] Murthy, CR; Rama Rao, KV; Baig, G; Norenberg, MD. Ammonia-induced production of free radicals in primary cultures of rat astrocytes. *J. Neurosci. Res.*, 2001, 66, 282–288.

[75] Nencki, M; Pavlov, IP; Zaleski, J. Ueber den Ammoniakgehalt des Blutes und der Organe. Die Harnstoffbildung bei den Saugetieren. *Arch. Exp. Path Pharmakol.*, 1896, 37, 26–51.

[76] Newsholm, E; Start, K. *Regulation in Metabolism*. Moscow:Mir; 1977. In Russian.

[77] Nishino, T; Okamoto, K; Eger, BT; Pai, EF; Nishino, T. Mammalian xanthine oxidoreductase — mechanism of transition from xanthine dehydrogenase to xanthine oxidase. *FEBS J.*, 2008, 275, 3278–3289.
[78] O'Connor, JE; Costell, M; Grisolía, S. Prevention of ammonia toxicity by L-carnitine: metabolic changes in brain. *Neurochem. Res.*, 1984, 9, 563–570.
[79] Ott, P; Clemmesen, O; Larsen, FS. Cerebral metabolic disturbances in the brain during acute liver failure: from hyperammonemia to energy failure and proteolysis. *Neurochem. Int.*, 2005, 47,13–18.
[80] Pan, G; O'Rourke, K; Dixit, VM. Caspase-9, Bcl-XL, and Apaf-1 form a ternary complex. *J. Biol. Chem.*, 1998, 273, 5841–5845.
[81] Panickar, KS; Jayakumar, AR; Rao, KV; Norenberg, MD. Ammonia-induced activation of p53 in cultured astrocytes: role in cell swelling and glutamate uptake. *Neurochem. Int.*, 2009, 55, 98–105.
[82] Petronilli, V; Penzo, D; Scorrano, L; Bernardi, P; Di Lisa, F. The mitochondrial permeability transition, release of cytochrome c and cell death. Correlation with the duration of pore openings in situ. *J. Biol. Chem.*, 2001, 276, 12030–12034.
[83] Petrosillo, G; Ruggiero, FM; Paradies, G. Role of reactive oxygen species and cardiolipin in the release of cytochrome c from mitochondria. *FASEB J*, 2003, 17, 2202–2208.
[84] Prives, C; Hall, PA. The p53 pathway. *J. Pathol.*, 1999, 187, 112–126.
[85] Qu, L; Huang, S; Baltzis, D; Rivas-Estilla, AM; Pluquet, O; Hatzoglou, M; Koumenis, C; Taya, Y; Yoshimura, A; Koromilas, AE. Endoplasmic reticulum stress induces p53 cytoplasmic localization and prevents p53-dependent apoptosis by a pathway involving glycogen synthase kinase-3beta. *Genes Dev.*, 2004, 18, 261–277.
[86] Qureshi, K; Rao, KV; Qureshi, IA. Differential inhibition by hyperammonemia of the electron transport chain enzymes in synaptosomes and non-synaptic mitochondria in ornithine transcarbamylase-deficient spf-mice: restoration by acetyl-L-carnitine. *Neurochem. Res.*, 1998, 23, 855–861.
[87] Rama Rao, KV; Jayakumar, AR; Norenberg, DM. Ammonia neurotoxicity: role of the mitochondrial permeability transition. *Metab Brain Dis.*, 2003, 18, 113–127.
[88] Rao, KV; Norenberg, MD. Cerebral energy metabolism in hepatic encephalopathy and hyperammonemia. *Metab. Brain Dis.*, 2001, 16, 67–78

[89] Ratnakumari, L; Murthy, Ch RK. Activities of pyruvate dehydrogenase, enzymes of citric acid cycle, and aminotransferases in the subcellular fractions of cerebral cortex in normal and hyperammonemic rats. *Neurochem. Res.*, 1989, 14, 221–228.

[90] Richter, C. Mitochondrial calcium transport. In: Ernster L editor. *Molecular Mechanisms in Bioenergetics*. Amsterdam: Elsevier; 1992; pp. 349–358.

[91] Rottenberg, H; Scarpa, A. Calcium uptake and membrane potential in mitochondria. *Biochemistry*, 1974, 13, 4811–4817.

[92] Rotter, V; Abutbul, H; Ben-Ze'ev, A. P53 transformation-related protein accumulates in the nucleus of transformed fibroblasts in association with the chromatin and is found in the cytoplasm of non-transformed fibroblasts. *EMBO J.*, 1983, 2, 1041–1047.

[93] Satoh, MS; Lindahl, T. Role of poly(ADP-ribose) formation in DNA repair. *Nature*, 1992, 356, 356–358.

[94] Satrustegui, J; Richter, C. The role of hydroperoxides as calcium release agents in rat brain mitochondria. *Arch. Biochem. Biophys.*, 1984, 233, 736–740.

[95] Scarlett, JL; Murphy, MP. Release of apoptogenic proteins from the mitochondrial intermembrane space during the mitochondrial permeability transition. *FEBS Lett.*, 1997, 418, 282–286.

[96] Schenker, S; McCandless, DW; Brophy, E; Lewis, MS. Studies on the intracerebral toxicity of ammonia. *J. Clin. Invest.*, 1967, 46, 838–848.

[97] Schraufstatter, IU; Hyslop, PA; Hinshaw, DB; Spragg, RG; Sklar, LA; Cochrane, CG. Hydrogen peroxide-induced injury ofcells and its prevention by inhibitors of poly(ADP-ribose)polymerase. *Proc. Natl. Acad. Sci. USA*, 1986, 83,4908–4912

[98] Shah, GM; Shah, RG; Poirier, GG. Different cleavage pattern for poly(ADP-ribose) polymerase during necrosis and apoptosis in HL-60 cells. *Biochem. Biophys. Res. Commun., 1996*, 229, 838–844.

[99] Sun, GY; Shelat, PB; Jensen, MB; He, Y; Sun, AY; Simonyi, A. Phospholipases A2 and inflammatory responses in the central nervous system. *Neuromolecular Med.*, 2010, 12, 133–148.

[100] Susin, SA; Lorenzo, HK; Zamzami, N; Marzo, I; Snow, BE; Brothers, GM; Mangion, J; Jacotot, E; Costantini, P; Loeffler, M; Larochette, N; Goodlett, DR; Aebersold, R; Siderovski, DP; Penninger, JM; Kroemer, G. Molecular characterization of mitochondrial apoptosis-inducing factor. *Nature*, 1999, 397(6718), 441–446.

[101] Van Caulaert, C; Deviller, C. Ammoniemie experimentale apres ingention de chlorure d'ammonium chez l'homme a l'etat normal et pathologique. *CR Soc. Biol. (Par)*, 1932, 3, 50.

[102] Van Caulaert, C; Deviller, C; Halff, M. Contribution a l'etude des cirrhoses de. Toxicite de l'ammoniaque. *Presse Med.*, 1933, 41, 217.

[103] Vaughan, H; Thornton, SD; Newsholme, EA. The effects of calcium ions on the activities of trehalase, hexokinase, phosphofructokinase, fructose diphosphatase and pyruvate kinase from various muscles. *Biochem. J.*, 1973, 132, 527–535.

[104] Venediktova, NI; Lopata, OV; Pogosyan, AS; Kosenko, EA; Kaminsky, YG. Antioxidant enzymes of rat liver, brain, heart and erythrocytes in ammonia intoxication. *Russ J. Biomed. Khim.*, 2005, 51, 185–191. In Russian.

[105] Venediktova, NI; Kosenko, EA; Kaminsky, YG. Antioxidant enzymes, hydrogen peroxide metabolism, and respiration in rat heart during experimental hyperammonemia. *Biol. Bull*, 2006, 33(3), 281–286.

[106] Vercesi, AE; Kowaltowski, AJ; Grijalba, MT; Meinicke, AR; Castilho, RF. The role of reactive oxygen species in mitochondrial permeability transition, *Biosci. Rep.*, 1997, 17, 43–52.

[107] Verkhratsky, A. Calcium and cell death. *Subcell Biochem.*, 2007, 45, 465–480.

[108] Whitacre, CM; Hashimoto, H; Tsai, M-L; Chatterjee, S; Berger, SJ; Berger, NA. Involvement of NAD-poly(ADP-ribose) metabolism in p53 regulation and its consequences. *Cancer Res.*, 1995, 55, 3697–3701.

[109] Worcel, A; Erecinska, M. Mechanism of inhibitory action of ammonia on the respiration of rat-liver mitochondria. *Biochim. Biophys. Acta*, 1962, 65, 27–33.

In: Ammonia: Structure, Biosynthesis... ISBN: 978-1-62100-502-5
Editors: V.A. Fekete, et al, pp. 33-60

Chapter 2

DEVELOPMENT OF DISTRIBUTED FIBER OPTIC SENSOR OF AMMONIA GAS

Ladislav Kalvoda, Jan Aubrecht and Petr Levinský
Czech Technical University in Prague.

ABSTRACT

Our contribution starts with a brief overview of the recent state-of-art in the field of 'classical' sensors routinely used in detection of ammonia gas. A short reference is made of a wide practical usage of ammonia gas and its harmful properties stimulating the ever-lasting emphasis on development of spatially continuous and highly sensitive sensors. Consecutive summary of alternative approaches taking advantage of utilization of optical fibers in place of the sensing element is then followed by a detailed theoretical treatment of the fiber optic distributed system employing the optical time domain reflectometry (OTDR) technique. The derived model is used in computer simulations, results of which are compared with the experimental data obtained in tests of real sensing fibers. Suggestions are then asserted concerning the promising directions of further development of the sensor system.

I. INTRODUCTION

The subject of our research, a distributed detection of ammonia gas performed by means of a fiber optic sensor, lies on intersection of two research areas that are very dynamically developing at present: the optical fiber-based sensors and the ammonia gas sensors. Thousands of research papers are published every year touching on problems within the thematic fields and hence, it is practically impossible provide a comprehensive overview of the obtained achievements referring to the original papers. With aim to somewhat simplify the task, if not particularly referenced, the brief summary given in the next part of this paragraph is mostly based on the recent review articles [1 - 12] and restricts preferentially on the sensing applications specifically designed for ammonia gas sensing. Thus, e.g. ammonium ion sensing techniques are not systematically referred here.

To start with, it is worth to notice that ammonia is known to be very likely playing an important role in the process of life forming on our Earth. During the past, the atmospheric ammonia concentration slowly decayed reaching the natural 10-100 ppb level over continents and sub-ppb levels observed above oceans at present. The main sources of ammonia in environment are recently related to human activities, the latter resulting in several tens of millions of tons of annual ammonia emission. Geographically, the maximum emission rate is located in the Western and Central Europe. There are three main groups of the contributing processes/activities: fixation of air nitrogen in soil (several percents of the total), domestic animal farming (the major part) and combustion processes in plants and motor vehicles (about 25 percent of the total).

The tabulated limit of human ammonia perception is around 50 ppm, corresponding to 40 μg/m3, but even below this limit ammonia is irritating to the respiratory system, skin and eyes. The long term allowed concentration that people may work in is set to be 20 ppm. Immediate and severe irritation of the nose and throat occurs at 500 ppm. Exposure to high ammonia concentrations, 1000 ppm or more, can cause pulmonary oedema. Extremely high concentrations, 5000–10,000 ppm, are suggested to be lethal within 5–10 min. Longer periods of exposure to low ammonia concentration are not believed to cause long-term health problems since ammonia is a natural body product excreted from the body in the form of urea and ammonium salts in urine and sweat.

II. Main Application Areas of Ammonia Sensors

For humans, a qualitative detection of ammonia at high concentrations is easy for sensitivity of human nose to ammonia is high. Nevertheless, we fail to recognize low ammonia concentrations as well as to quantify the high concentrations. In such cases, application of artificial sensing device is necessary. Roughly say, four application areas can be distinguished demanding either a quantitative or a highly sensitive detection of ammonia: environmental, automotive, chemical industry and medical diagnostics.

II.1. Environmental Ammonia Sensors

In farming areas or sites with heavy automotive traffic, the ammonia concentration can reach several tens of ppm and the smell makes life unpleasant. Locally, e.g. in stables or dung-yards, ammonia concentrations can even exceed the safety limits. Another effect of large atmospheric ammonia concentration is forming of ammonia sulfate and nitrate aerosols acting as condensation centers and contributing to smog formation. The detectors intended for open-air environmental ammonia analysis are required to provide a high sensitivity down to ppb range, but no fast response is needed. The detectors intended for use in stables or other closed areas have to provide sensitivity down to ppm range and response time in range of minutes.

II.2. Automotive Ammonia Sensors

Applications of ammonia sensors related to automotive industry include detection of ammonia in the car exhaust (optimization of the engine operation, catalytic conversion of NO_x, reduction of the related airborne aerosols; low detection limit in range of ppm, detection time in sub-second range, high temperature stability up to 1000 ^{o}C required) and registration of ammonia in the air blowing into car compartment (detection limit about the safety limit 50 ppm, detection time in range of seconds).

II.3. Industrial Ammonia Sensors

In chemical industry, ammonia is produced by catalyzed Bosh-Haber process and used mainly in production of fertilizers, chemicals and as a coolant medium in large-scale refrigerants used for food processing, storage or in ice-hockey hall technologies. The latter usage is stimulated by the ability of ammonia to cool below 0 ◦C. In al the cases, pure ammonia is used and so very dangerous situation occurs in case of leak. The detection systems must be in this case featuring sensitivity threshold at 20 ppm (allowed ammonia concentration), response time in seconds and provide a high spatial coverage of the controlled area. In case of ammonia production plants, an enhanced temperature stability of the sensors can be required.

II.4. Medical Ammonia Sensors

Concerning the medical applications of ammonia sensors, the most important part relates to the breath gas analysis (BGA). Breath gas analysis (BGA) has a high potential for diagnosis of physiological state of humans. Over 1000 volatile compounds are contained in ~ 0.9% (w/v) of total amount of each human breath. One of them is ammonia. The primary metabolic product of food 'processing' in mammals is urea, produced in livers. Ammonia plays important role in this process as a 'raw' material as well as a product. Together with urea, it is subsequently filtrated out from bloodstream in kidneys. In case of kidneys malfunction, concentration of ammonia in blood exceeds the normal limit - ca 6.6 ppb. The high ammonia blood concentration can then result in brain damage (hepatic encephalopathy, in later stages in type II Alzheimer disease). In cases when ammonia concentration in blood is higher then the ambient concentration, ammonia diffuses from blood to lungs and is breathed out. The latter process is instrumental for a non-invasive analysis of ammonia content in blood. The normal physiological range of ammonia is in the region of 50 to 2,000 ppb what defines the required sensing sensitivity limit. A high selectivity to ammonia in presence of carbon dioxide and response time in range of minutes is required, too. Skipping the laboratory methods of BGA, the recent development is focused on so-called point-of-care monitors providing immediate information.

There are several important medical applications related to analysis of ammonia concentration in breath gas [9]: (i) non-invasive observation of haemodialysis process efficiency (detection of efficacy of urea extraction from blood), (ii) analysis of peptic ulcers of stomach or duodenum caused by bacteria Helicobacter pylori (if H. pylori is active in the body, ingestion of urea leads to significantly higher increase of ammonia content in breath gas ~ 400 ppb - then in the opposite case ~ 150 ppb; the latter difference is caused by urease produced by the bacteria), (iii) there is relationship between asthma appearance and the ammonia breath concentration - persons with asthma show lower ammonia concentration in breath then healthy individuals, (iv) there are bacteria that can generate ammonia in oral cavity - about 90% of breath odor originates as a result of orolaryngeal and/or gastrointestinal disorders - halitosis, (v) the expired ammonia levels increase exponentially with the body workload - the ammonia concentration obtained by BGA can be also used to monitor the load intensity during a sport activity; the concentration levels of interest are in the range of 0.1 to 10 ppm.

III. Principles of Ammonia Sensors

The most common ammonia sensing principles investigated at present include metal-oxide sensors, catalytic sensors, conducting polymer sensors, sensors based on nano-sized structures and optical sensors. In the following, we briefly summarize their principle characteristic features.

III.1. Metal Oxide Sensors

Ammonia sensors based on metal oxide layers (tin dioxide being the most common) are the most widely used ammonia sensors today. Surface electric conductivity of the layer is measured and correlated to the concentration of adsorbed ammonia. Detection limits of these sensors are in concentration range 1 - 1000 ppm. The sensing layer has to be kept at an elevated temperature (400 - 500 °C) ensuring that an efficient desorption of ammonia molecules necessary for the sensor reversibility takes place. Main drawbacks of the sensing elements include detection cross-sensitivity, insufficient long-term stability and relatively high electric power consumption.

III.2. Catalytic Sensors

In catalytic sensors, the charge carrier concentration in the catalytic metal depends on concentration of the target gas and can be quantified using a field effect device, such as capacitor or FET. Ammonia FETs with a palladium gate material have been prepared showing a detection sensitivity threshold 1 ppm. If a capacitor configuration is used in combination with an ion-conducting electrolyte, the electric potential difference of the resulting cell directly relates to the chemical concentration of the target gas. The low sensitivity limit of such chemical cells is in range of ppm.

III.3. Conductive Polymer Sensors

Two conductive polymeric materials have been reported as used in construction of ammonia sensors: polypyrrole and polyaniline. In case of polypyrrole, two reactions mechanisms are possible: irreversible ammonia addition to the polymer chain or reversible reduction of oxidized form of the polymer. Using a thin film of the polymer, course of the former reaction can be detected e.g. by quartz microbalance resonator and extend of the latter reaction analyzed by surface electric conductivity measurements. Stability of the sensors prepared with polyaniline layers proved to be superior compared to polypyrrole ones. Ammonia is believed to cause deprotonation of polyaniline leading to electric conductivity and optical absorbance changes of the polymer. Thus, both electric and optical methods can be used to read out the sensor. Polypyrrole, polyaniline and hydroquinone films were also tested as ammonia sensors interrogated by electro-chemical methods [4].

III.4. Sensors with Nano-Sized Structures

Nano-patterned structures provide a new promising direction in development of ammonia sensing heads. Physical methods of sensor nano-materials preparation include different versions of vacuum deposition. High reactivity of nano-materials at low temperatures has led to appearance of cryochemistry - a new field of chemistry dealing with synthesis of nano-structures at low temperatures. Examples of nano-structures successfully applied in construction of selective ammonia sensors involve titanium dioxide covered by carbon nanotubes [13], structures based on carbon nanotubes sensitized with polyaniline [14] or nafion [15] polymer, lead nanoparticles [16], silver oxide [17], vanadium oxide [18] and zinc oxide [19] nanowires.

III.5. Extrinsic Optical Sensors

As extrinsic optical sensors for ammonia detection we denote here systems taking advantage of one of the following two sensing principles: (i) detection of absorption band changes accompanying a reaction of ammonia with the selected reagent (dye) or (ii) direct registration of intensity of some of the near-infra red (NIR) or mid-infra red (MIR) ammonia molecular absorption bands.

The former method can be implemented as spectrometric detection of the coloring reaction of ammonia gas when dissolved in aqueous solution of the reagent (such as Nessler or Berthelot reaction) or combined with application highly sensitive detection method, such as single photon counting. Lower specificity of the sensing reaction can be overcome by application of a selective membrane used to filter the stream of the analyzed gas.

One of the main methods used to realize the measurements of the NIR and MIR gas spectra is a photo-acoustic spectroscopy (PAS). It provides the most sensitive, ready to use tool for ammonia detection at present, with the lower sensing limit in range of ppb. Main disadvantage rendering application of this technique is the relatively complicated equipment including quantum cascade laser or NIR laser diode and acoustic resonator cell(s) necessary to record the gas absorption spectra [5, 6]. Another approach uses a radiation source with a wavelength near 2 μm (preferably 1993 nm) to measure the presence of ammonia, CO2 and water vapor simultaneously [20]. Employment of a pressure near 100 Torr decreases broadening of the different spectroscopic transitions, thereby isolating the corresponding absorption lines and enabling specific measurements of each analyte without interference.

III.6. Intrinsic Optical Sensors

Under this group we may consider mainly the optical waveguide sensors utilizing evanescent field interaction of the guided light modes with the surrounding medium to collect the sensor signal. Both single- or multi-mode waveguide and planar or circular waveguide geometry can be used, in special cases further modified by local bends, branches etc. For instance, in case of ammonia, fiber optic sensor was reported in [7] the working principle of which is based on active cladding prepared from polyaniline. The polymer can be switched to oxidized (electrically conducting) state by exposition to HCl and then reduced back to insulating state by contact with ammonia. The oxidized state of PA emeraldine base shows optical absorption maximum at ca 650 nm, the reaction with ammonia results in the red shift to ca 780 nm. The transition is also accompanied by RI (RI) change [11]. Related changes in the light energy absorbed in the cladding are trough evanescent field components of the individual modes transferred to the guided modes, resulting in total changes of the guided light intensity that is registered. Detection efficiency of such intrinsic fiber optic sensor (FOS) built up on a multi-mode optical fiber can be further enhanced by application of modal power distribution (MPD) measurement technique. Further variations of the described principle can be obtained by application of a properly selected reagent changing its fluorescence intensity or RI in presence of the target analyte.

Efficient, low cost, sensitive and fast response intrinsic evanescent wave ammonia sensing heads could be also prepared by application of sol-gel technique. Lobnik and Wolfbeis [21] developed an optical sensor for the continuous determination of dissolved ammonia, by incorporating aminofluorescein in organically modified sol–gel films prepared by co-polymerisation of TMOS and diphenyldimethoxysilane. The dynamic range was from 1 to 20 ppm and storage stability (in distilled water) was over 6 months. Malins et al. [22] proposed a compact planar waveguide ammonia sensor for personal monitoring tasks in industrial environments, based on a cyanine dye doped sol–gel film; this sensor was fully reversible, presented limit of detection of 5 ppm and response time of a few seconds. MacCraith and co-workers [23] and, more recently, Cao and Duan [24] as well as Tao et al. [25], used sol–gel films doped with bromocresol purple and coated on unclad optical fibers for detecting gaseous ammonia. Silica sol-gel doped with silver nanoparticles and coated on an optical fiber allows ammonia sensing with the low sensitivity threshold in sub-ppm range [26]. Exposure of the nanocomposite-coated bent optical fiber probe to a gas containing ammonia reversibly enhances the power attenuation of the light guided through the U-shaped optrode.

Another common intrinsic fiber optic sensor configuration suitable for remote sensing is the single-ended reflection optrode. For instance, detection of trace ammonia can be done by rapidly swept continuous wave cavity ringdown spectroscopy [27]. Measurements in the NIR range of 1.51-1.56 μm yield ppb or better sensitivity in the gas phase for several representative gases (CO_2, CO, H_2O, NH_3, C_2H_2 and other hydrocarbons). Thin films of zirconia (ZrO_2) nanoclusters and poly(sodium 4-styrenesulfonate) salt deposited on the cleaved ends of telecommunication optical fibers are able to operate as ammonia sensors under ambient conditions without heaters, and show zero or negligible cross-sensitivity to humidity, temperature and volatile organic compounds [28]. Specially tapered and polished optical fibers were used to directly detect ammonia by modification of SPR conditions. The main disadvantage of this device was lack of selectivity and laborious optrode fabrication [29]. Stimuli-sensitive gelatin films containing photochromic bacteriorhodopsin nanofragments from Halobacterium salinarum at the distal end of an optical fiber were used to reversibly detect ammonia or water vapors through a color change [30]. Ability of a minimally invasive, highly sensitive sensor (comprised of optical fiber, the distal end of which consists of a pH-sensitive colorimetric dye embedded in a gas-permeable layer) to detect ammonia in the breath of patients with end-stage liver disease was published [31].

Gas-filled photonic band-gap fibers (PBF; 1–10 m long) are also tested as promising sensing principle for highly sensitive ammonia detection. Ritari et al [32] used PBF filled with ammonia gas to record ammonia absorption spectrum within the NIR range 1300–1600 nm in order to determine ammonia concentration. The principal drawbacks found during the experiments involved complexity associated with filling/evacuating the PBF with the target gas and a strong adsorption of ammonia onto the silica surface of the PBF restricting reversibility of the detection system.

III.7. Distributed Sensing Systems

Standard architecture of such systems used at present is based on a network of individual sensing heads, prevailingly employing some type of the metal oxide-based sensing elements. Installation of distributed networks of ammonia sensors is necessary wherever danger of massive ammonia leak exists, especially in closed facilities using large-scale refrigerating facilities, such as abattoirs, dairies, canneries, refrigerator vessels, shopping malls, skating rings and ice hockey halls, or industrial plants producing and/or storing large volumes of ammonia gas. The sensing network provides the tool for early warning when the ammonia leak occurs necessary for securing and evacuation of the affected area. Such sensor network could be also very instrumental in farming facilities with a permanent ammonia emission (such as stables or henneries) for it can provide early warning in case of a venting system failure. It is obvious that cost, electric power consumption and operational complexity of such sensing head network rapidly grows as number of the installed sensor heads rises up. Moreover, application of discrete sensing elements can never ensure complete coverage of the inspected space.

Numerous sensing schemes and optical systems have been tested to overcome at least some of the mentioned drawbacks (e.g. [33 - 38]). One of the promising approaches employs the already mentioned intrinsic FOS concept based on fiber cladding sensitization with a proper reagent showing change of some of its optical properties (spectral changes of its complex RI or fluorescence emission) in presence of the target analyte - ammonia gas. Various modifications of optical reflectometry approach (optical time domain reflectometry - OTDR, optical frequency domain reflectometry - OFDR, optical low coherence reflectometry - OLCR, optical time-of-flight chemical detection - OTOF-CD) can be then employed in construction of resulting distributed sensing systems [39 - 41].

IV. Distributed Fiber Optic Sensor of Ammonia with OTDR Readout

The ammonia detection principle employed in construction of our sensing fiber utilizes a ligand exchange reaction proposed in [42] (see formula (1) below). The principal asset of this reaction schema compared with, for instance, an acid-base reagent such as bromocresol purple is in the high specificity of the reaction towards ammonia.

The system consists of two main parts: (i) the sensing optical fiber working as an intrinsic chemo-optical absorption-based transducer and (ii) an interrogating OTDR unit (Figure 1). A step-index, multimode, plastic clad silica (PCS) fiber was used in our experiments with the core radius a = 0.01 cm, the polysiloxane cladding radius b = 0.012 cm, the core RI n_0 = 1.465, the cladding RI n_1 = 1.455, the numerical aperture $NA = (n_0^2 - n_1^2)^{1/2}$ = 0.171 and the critical angle θ_c = 6.7 deg. The cladding is supposed to be sensitized by an ammonia-sensitive reagent R possessing a particle density N^R [cm^{-3}], molecular extinction coefficient ε^R [cm^{-1}] and molecular polarizability δ^R [cm^3].

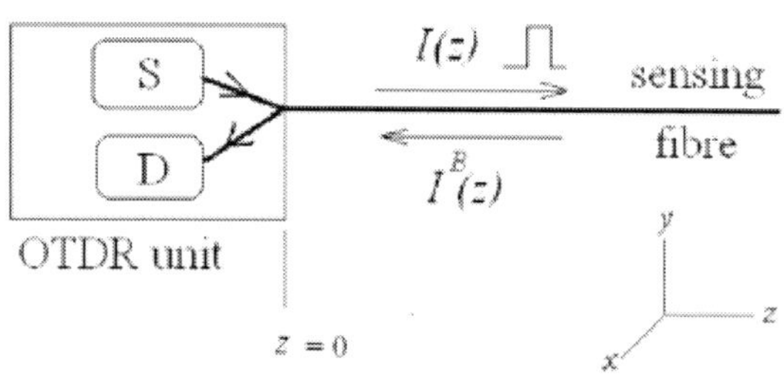

Figure 1. Schema of the sensing system and definition of the reference co-ordinate system. S – the pulsing laser diode, D – the light detector, I – the light intensity propagating along the +z direction, I^B – the intensity propagating along the -z direction.

The sensing process includes two steps: (i) diffusion of ammonia (the analyte) into the cladding and (ii) conversion of the reagent in presence of the analyte accompanied by spectral change (Figure 2). The conversion reaction is supposed to be fully reversible, concentration-controlled composition/decomposition of the organo-metallic complex reagent R = $(L_n\text{-}Me)^{p+}$ $(A^-)_p$ (L – organic ligand, n – number of ligands in the reagent molecule, m – number of ligands in ammonia complex, Me – the central metal ion, p – the oxidation state of Me, A – a selected univalent counter-anion):

$$(L_n\text{-}Me)^{p+} + (A^-)_p + m(NH_3) \leftrightarrow ((NH_3)_m\text{-}Me)^{p+} + (A^-)_p + nL \qquad (1)$$

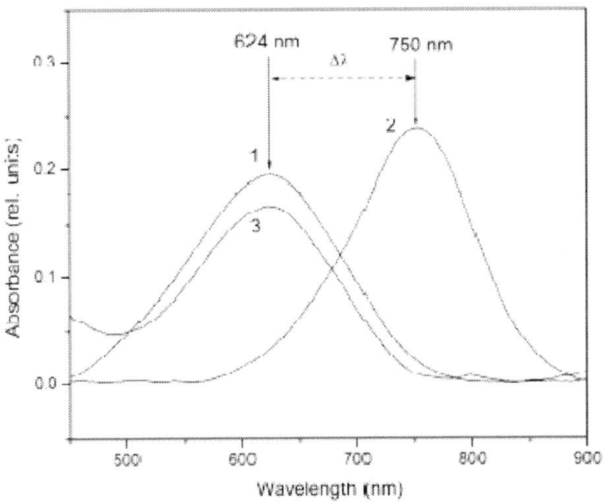

Figure 2. Example of spectral changes of VIS optical absorption following the chemical reaction (1) when Me = Cu, A = SO4 and L = 5-(4'-dimethyl amino phenylimino) quinolin-8-one. Spectrum 1 - the ligand L; 2 - the complex reagent R; 3 – the spectrum after ammonia liquor addition. The bathochromic shift Δλ is indicated. Spectra recorded in ethanol.

The OTDR unit launches a short rectangular light pulse into the fiber. For simplicity, we neglect the frequency dispersion of the pulse and consider it as monochromatic radiation. Thus, the pulse is characterized by its irradiance $I(z=0) = I_0$ [W cm^{-2}], duration τ [s], spatial width $w = \tau c/n_0$ [cm] and wavelength λ [cm^{-1}]. For sake of simplicity and due to the fact that a weakly guiding fiber is used, the real modal structure of the electro-magnetic (EM) waves guided within the fiber core (each mode generally characterized by the azimuth index l and radial index m pair (l,m) [43]) is approximated by the first mode $(l,m) = (0,1)$. Furthermore, the power density distribution across the core is supposed to be uniform, with the value $I^{core}(z)$ independent of the fiber radius $r = (x^2 + y^2)^{1/2}$ and the azimuth angle $\varphi = arctan(y/x)$. The parameter $\gamma_{(0,1)} = \gamma$ [cm^{-1}] characterizing the decay of the radial evanescent component of the power density within the fiber cladding is then obtained as the solution to the characteristic equation [43]

$$X\frac{J_1(X)}{J_0(X)} = Y\frac{K_1(Y)}{K_0(X)} \tag{2}$$

For X and Y holds

$$X = a\left\{\left(\frac{2\pi NA}{\lambda}\right)^2 - \gamma^2\right\}^{1/2}, \quad Y = \gamma a \tag{3}$$

Functions J_l and K_l represent the Bessel functions of the first kind and the modified Bessel functions of the second kind, respectively. For the above specified fiber we obtain $\gamma = 1.26\ \mu m^{-1}$.

The primary pulse experiences several interactions as it propagates along the fiber ($+z$ direction): the Rayleigh scattering occurring on the core and the cladding matrix materials, and the absorption and Rayleigh scattering in contact with molecules of the reagent. All the processes reduce the primary pulse intensity. Progressive interference of spherical waves created at individual scattering centers is then giving rise to two planar waves propagating along the $+z$ and $-z$ directions. The first wave adds to the primary pulse and its effect can be omitted. Time evolution of the intensity of the back-propagating wave ($I^B(t)$) is registered at the OTDR unit detector. The temporal co-ordinate is converted to the spatial position along the fiber: $z = ct/(2n_0)$. The typical $I^B(t)$ course (Figure 3) contains two Fresnel reflections originating at both ends of the tested fiber and an intermediate part providing us with information about optical properties along the tested fiber length.

Reagent molecules are supposed to be distributed within the cladding polymer with the mass concentration $u^R(r,z,t)$ (M^L and M^R marks the molecular weight of the ligand the reagent, respectively; N_A is the Avogadro number)

$$u^R(r,z,t) = (M^R/N_A)\, N^R(r,z,t) \tag{4}$$

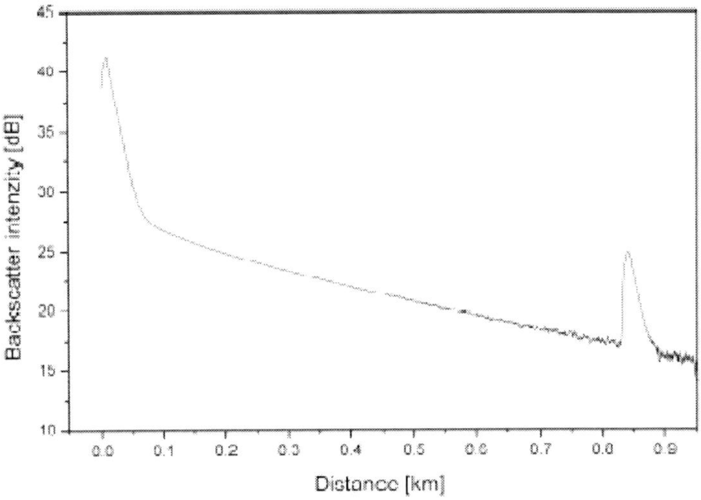

Figure 3. Example of OTDR signal recorded on a multimode unbuffered PCS fiber (length 832 m, b/a = 120 μm /100 μm); OTDR unit Agilent E6000C/E6005A (τ = 100 ns, λ = 850 nm).

Let's now expose the fiber to ammonia within the interval $z_1 \leq z \leq z_2$. Obviously, the mass concentration of ammonia in the cladding volume ($u^A(r,z,t)$, $r \leq b$, $t>0$) will start to grow according to the second Fick's law

$$\frac{\partial u^A}{\partial t} = \frac{\partial}{\partial r}\left\{ D\left(\frac{\partial u^A}{\partial r}\right)\right\} - k^{AR}u^R u^A \qquad (5)$$

For sake of simplicity, we will in further suppose that (i) the coefficient of ammonia solubility in the cladding is equal to unity, i.e. the concentration $u^A(r=b-\Delta r) = u^A(r=b+\Delta r)$ for any $\Delta r>0$, and (ii) the diffusion coefficient of ammonia gas (D) in the cladding does is constant. The second term on the right side of Equation (5) quantifies the trapping process accompanying the chemical reaction of ammonia with reagent. The parameter k^{AR} [$cm^3g^{-1}s^{-1}$] is the speed constant related to the first order reaction between ammonia and reagent: A + R ↔ AR adopted as approximation of the real reaction course (1). Apparently, the radial concentration of reagent $u^R(r,z,t)$ will also change during the exposition:

$$\frac{\partial u^R}{\partial t} = -q u^A u^R \tag{6}$$

Boundary conditions of the processes described by Equations (5) and (6) are given as

$$u^A(r < b, t = 0) = 0 \tag{7a}$$

$$u^A(r = b, t \geq 0) = u_0^A = const \tag{7b}$$

$$u^R(a \leq r \leq b, t = 0) = u_0^R = const\,. \tag{7c}$$

The differential equations (5) and (6) together with the boundary conditions (7) can be transformed into the following difference equations by setting $t = i\,\Delta t$, $r = a + j\,\Delta r$; $\Delta r = a + j\,(b - a)/100$; $i, j \in N_0, j = 0, .. 100$:[1]

$$u^{A,i+1}_{\ j} = u^{A,i}_{\ j} + \frac{\Delta t D}{(\Delta r)^2}(u^{A,i}_{\ j+1} - 2u^{A,i}_{\ j} + u^{A,i}_{\ j-1}) - q\Delta t u^{R,i}_{\ j} u^{A,i}_{\ j} \tag{8}$$

$$u^{R,i+1}_{\ j} = (1 - q\Delta t u^{A,i}_{\ j}) u^{R,i}_{\ j} \tag{9}$$

$$u^{A,i=0}_{\ j} = 0, \;\; j = 0,..100 \tag{10a}$$

$$u^{A,i}_{\ j=0} = u_0^A, \;\; i \geq 0 \tag{10b}$$

$$u^{R,i=0}_{\ j} = u_0^R, \;\; j = 0,..100 \tag{10c}$$

The derived explicit difference schema with the precision of order $O(\Delta t + \Delta r^2)$ is numerically stable if the temporal (Δt) and spatial (Δr) step satisfy the Courant's stability condition

[1] Further on, we suppose that the radial co-ordinate is discretized into 100 intervals.

$$D\frac{\Delta t}{(\Delta r)^2} < \frac{1}{2} \tag{11}$$

Equations (8) – (11) allow for calculation of the sought radial and temporal evolution of the ammonia and reagent concentrations using a simple iteration schema. Starting with $t = 0$ ($i = 0$), the concentration profiles of ammonia and reagent are set according to (10) and the time step Δt set to agree with (11). Then, the vector $u_j^{A,i=1}$ is obtained from (8) and using (9) we get the new $u_j^{R,i=1}$. The latter vector enters Relation (8), i is incremented and the process is repeated.

Let's now quantify the power of the primary light pulse propagating along the fiber. In order to evaluate the interaction between the interrogating light pulse and the fiber matter, we need to know (i) the current radial concentration profile of reagent and (ii) the values of several important material constants: the molecular extinction coefficient (ε^R) and the molecular polarizability (δ^R) of the reagent and the loss coefficient (α^M) and the average molecular polarizability (δ^M) of the core and the cladding material[2].

Total light power $I(z)$ [W] at position z can be written as

$$I(z) = I_1(z) + I_2(z) \tag{12}$$

$$I_1(z) = \pi a^2 I^{core}(z) \tag{13}$$

$$I_2(z) = 2\pi I^{core}(z)\int_a^b \exp\{-2\gamma(r-a)\}rdr \tag{14}$$

The subscript "1" and "2" refers here to the fiber core and cladding, respectively. Approximation of the Bessel functions $K_l(\gamma r)$ (giving the correct radial decay of the evanescent electric field in the cladding) by a simple exponential function is used in Equation (14). The corresponding integral can be then solved analytically. The mean irradiance ($I^{core}(z)$, [W cm^{-2}]) within the fiber core is introduced as

[2]For sake of simplicity, we suppose in further that the molecular polarizability is identical for both matrices. Both fused silica and polysiloxane resin matrix contains extended siloxane network, though differing in density and topology.

$$I^{core}(z) = \frac{I_1(z)+I_2(z)}{\pi}\left\{a^2 + \frac{2}{\gamma^2}\left(1-\exp(-\gamma(b-a))(\gamma(b-a)+1)\right)\right\}^{-1} \quad (15)$$

Light power transported between two points z_1 and z_2 ($z_1<z_2$) is damped. The damping is different within the core and the cladding cross-section:

$$I_1(z_2) = I_1(z_1)\exp\left(-\alpha^M(z_2-z_1)\right) \quad (16)$$

$$I_2(z_2) = I_2(z_1)\exp\left(-(\alpha^M+\alpha^R(z_1 \leftrightarrow z_2))(z_2-z_1)\right) \quad (17)$$

The linear absorption coefficient $\alpha^R(z_1 \leftrightarrow z_2)$ averaged over the fiber cladding within the interval $\langle z_1,z_2 \rangle$ can be calculated from the corresponding reagent radial concentration profile weighted by the evanescent field intensity

$$\alpha^R(z_1 \leftrightarrow z_2) = \frac{2\pi N_A \ln(10)\varepsilon^R}{M^R}\int_{z1}^{z2}\int_a^b u^R(r,z)\exp(-2\gamma(r-a))rdrdz \quad (18)$$

The factor ln(10) in (18) respects the transformation from the decimal logarithmic base used in definition of ε^R to the natural logarithm applied in (17). The integral in Equation (17) can be solved numerically using a trapezoidal integration schema.

In order to simulate the intensity profile of the light propagating along the fiber the axial co-ordinate (z), finite difference grid is introduced characterized by the step size ($\Delta z = z_2 - z_1$).[3] At every nexus of the grid, the components I_1 and I_2 of the intensity coming from the previous node are determined using Equations (16) ÷ (18) and the total core intensity is obtained from (15). The new components I_1' and I_2' are then calculated from Equations (13) and (14) and used in evaluation of the intensity coming to the next nexus. Etc. The re-distribution of the intensity between the core and the cladding ($I_1 \rightarrow I_1'$, $I_2 \rightarrow I_2'$) reconstructs the necessary intensity balance distorted primarily by the reagent absorption within the cladding layer. This schema can be also

[3] Proper step size must be selected to ensure that the intensity $I_2(z_2)$ in (17) does not completely diminish after light pulse propagation over the step width.

understood as a discrete application of the conditions of continuity binding together the corresponding EM field intensities at the core/cladding boundary.

Intensity of a back-scattered wave $I^B(z_1)$ arisen due to the Rayleigh scattering at the position z_1 is given as

$$I_1^B(z_1) = I_1(z_1)R_1 \tag{19}$$

$$I_2^B(z_1) = I_2(z_1)R_2 \tag{20}$$

For the light intensity propagating within the fiber cladding is about 1 percent of the total transmitted intensity, the contribution resulting from the Rayleigh scattering occurring in the cladding can be neglected and the coefficient R_2 set $R_2 = 0$. The coefficient R_1 in Equation (19) corresponding to the back-scattered intensity originating in the fiber core can be expressed as

$$R_1 = \pi a^2 N^M \Delta z\left(S^B \frac{NA^2}{2n_0^{\ 2}}\right)\left(\frac{2\sqrt{3}\pi^{5/2}\delta^M}{\lambda^2 d}\right)^2 \tag{22}$$

The left-most term on the right side of Equation (22) provides the total of scattering centers within one ‘spatial cell’ of the width Δz. The second term gives the back-scattered portion of the scattered intensity re-bound into the fiber core and propagating along the $-z$ direction. The parameter S^B quantifies the overall efficiency of the back-scattering process; we set $S^B \equiv ½$. The third term is an integral efficiency of a single Rayleigh scattering source ‘observed’ from the distance d [44]. The later parameter is approximated by the path between a scattering point 1 lying on the axis z and an ‘observation’ point 2 on the core/cladding boundary ‘hit’ by a ‘representative’ ray propagating from the point 1 to the point 2 at the angle $\theta = \theta_c/2$ with respect to the axis z. Using the current fiber parameters n_0, n_1 and a, we get $d \approx 1.7$ mm.

Transport of the back-scattered light intensity back to the detector is again governed by Relations (12) - (17); the necessary re-distribution of the intensities I^B_1 and I^B_2 has to be again applied analogically to the model used for the light intensity propagating in the $+z$ direction.

V. Simulated Results and Their Comparison with Experimental Data

The theoretical model described in the former chapter has been transferred in form of computer code and used for simulation performed with aim to optimize fabrication and operation of a real sensing system. The first basic findings obtained with the model were published in [45] and can be briefly summed up in the following four recommendations for achievement of fast and well resolved sensing system response when an optical absorption-based intrinsic sensing fiber is used:

1) It is very desirable to keep the thickness of the fiber cladding as thin as possible and, in the same time, maximize the diffusion coefficient D. However, the issues related to mechanical robustness of the fiber must be carefully considered in the same time.
2) Usage of a long-wavelength light source together with a small core radius and a small numerical aperture of the sensing fiber enhance the evanescent interaction between the core and the cladding.
3) Application of the reagents is desirable featuring a selective chemical reaction with ammonia characterized by a high reaction speed constant k^{AR}.
4) Preparation of fibers with an irregular distribution of reagent concentrated on the external side of the cladding can be used to increase the sensing reaction at the simultaneous preservation of the fiber transmittance.

Optical fiber with the same parameters as upper mentioned was employed in the calculations. The other parameters of the system were then as follows: the fiber cladding layer was supposed to be doped with an organo-metallic reagent featuring (cf. Equation (1)) L = 5-(4'-dimethylaminophenylimino)quinolin-8-one, Me = Cu, A = Br, $p = 2$, $n = 2$ and $m = 4$. The molecular weight of the ligand and reagent was $M^L = 277.38$ g/mol and $M^R = 773.5$ g/mol, respectively. Because of lack of relevant experimental data, value of the diffusion coefficient of ammonia in a polysiloxane was at least roughly estimated from analogy with diffusion of molecular oxygen measured in [46]. Here, $D = 2\times10^{-10}$ cm^2/s was obtained for molecular oxygen diffusing through a Sylgard 184 cross-linked matrix[4]. Taking in account the fact that NH_3 molecule is slightly smaller in size then O_2 and neglecting possible Columbic and polarization effects, we supposed that $D \cong 10^{-9}$ cm^2/s in case of ammonia. The molecular extinction coefficient of the reagent R determined from optical spectra measurements in solution was $\varepsilon^R(\lambda = 850$ nm) $= 1.5\times10^{-17}$ cm^{-1}. Polarizability of the core and cladding matrices was approximated by the polarizability of SiO_2 molecule $\delta^M = 2.86\times10^{-24}$ cm^3 [47]. Corresponding average concentration of scattering centers was then calculated from the mass density of quartz ($\rho^{SiO2} \cong 2.2$ g/cm^3) and its molecular weight (M^{SiO2}) as $N^M = N_A\rho^{SiO2}/M^{SiO2} \cong 2.21\times10^{22}$ cm^{-3}. The intrinsic loss of the primary pulse intensity was characterized by the mean loss coefficient $\alpha^M = 6.9\times10^{-6}$ cm^{-1} corresponding to the total loss ~3 dB/km.

The next series of simulations presented here results in more specific predictions and recommendations. In the following we discuss several examples of the obtained findings.

The first groups of results relates to the effects linked to the possibility of an inhomogeneous radial distribution of reagent. As already mentioned above, the distribution of reagent with its concentration decreasing towards the fiber core provides a higher sensor response at the simultaneous preservation of the fiber transmission loss (Figure 4). The latter fact is important for a higher loss relates to a higher OTDR signal noise and reduces thus the overall signal-to-noise ratio of the sensing system.

From Figure 4 we see that the sensor signal is more then doubled in case of the linearly varying reagent distribution. Even higher increase is obtained for a parabolic profile shape. This observation leads us to conclusion that the reagent molecules located in vicinity of the fiber core have a non-desirable, parasitic influence on the sensor function: (i) the analyte needs a longer time to diffuse in this region and (ii) owing to the strong evanescent field interaction in this region, the faster but weaker response from the positions in the cladding more distant from the core is suppressed.

The effect is even more pronounced when a highly spatially localized, rectangular radial concentration profile (2 μm in width) is considered and let to move step by step away from the core (Figure 5). It is apparent that for both the tested reaction speed constants, $k^{AR} = 0.5$ and 1000 cm^3 g^{-1} s^{-1}, the optimal radial position exist that maximizes the response to ammonia exposition.

[4] Sylgard (Dow Corning) is a commercial thermo-cross-linkable fiber cladding material based on polysiloxane formulation.

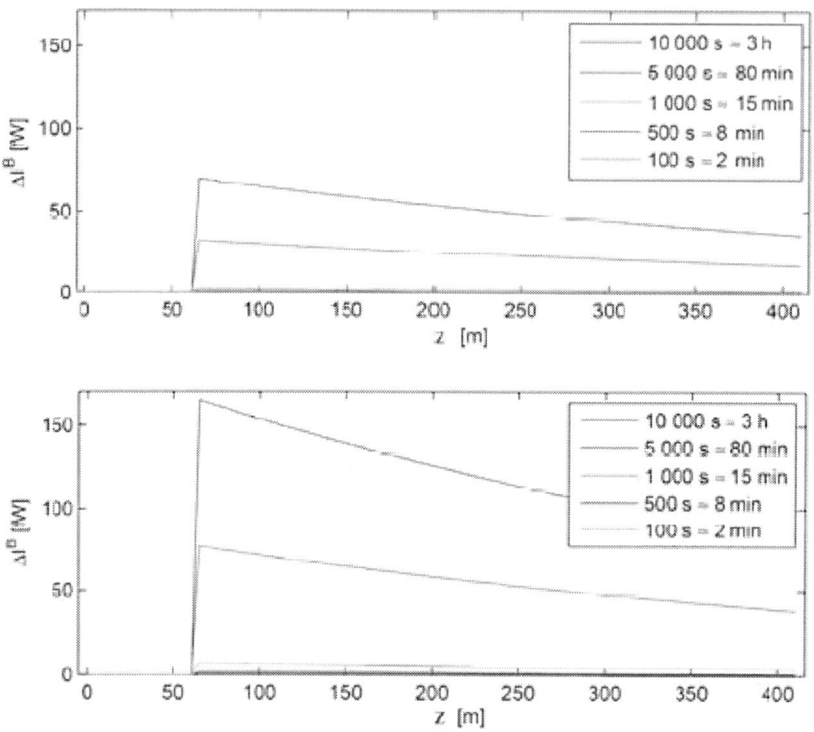

Figure 4. Change of the back-scattered intensity $\Delta I^B(z)$ after local exposition with ammonia calculated for two different types of the reagent radial concentration profile and 5 different exposition periods. The fiber transmission loss kept constant at the level 4.2 dB. On the left - a constant concentration profile; on the right - a profile linearly descending towards the core. Other parameters: $D = 10^{-9}$ cm^2/s, $v_0 = 5000$ ppm, $k^{AR} = 0.5$.

Analogical result is obtained when the position of the doped region is fixed (the middle 2 μm distant from the core/cladding boundary) and only the reagent concentration is changed (Figure 6). Again, depending on the current reaction speed constant, there is an optimum reagent concentration maximizing the response. So, in no case is truth that the maximal reagent concentration leads to the utmost OTDR response signal! In case of a too high reagent concentration, most of ammonia diffusing into the cladding is bound to form ammonia-metal complex in far distance from the fiber core where the evanescent field is weak, thus resulting in only small sensor response occurring for short registration periods or low analyte concentrations.

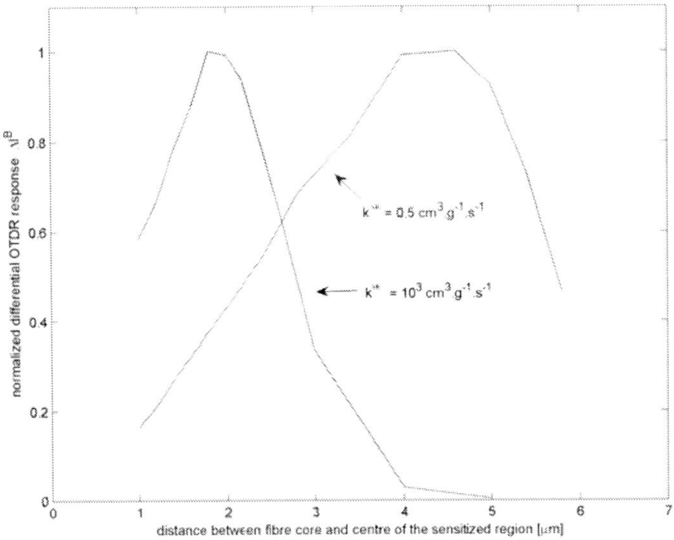

Figure 5. Dependence of the ammonia-induced $\Delta I^{B}(z)$ on the middle position of a 2 µm wide cladding zone with uniform reagent concentration, calculated for the two indicated reaction speed constants. Other parameters: $D = 10^{-9}$ cm^2/s, $v_0 = 5000$ ppm, exposition time 1000 s, fiber loss 4.2 dB kept constant.

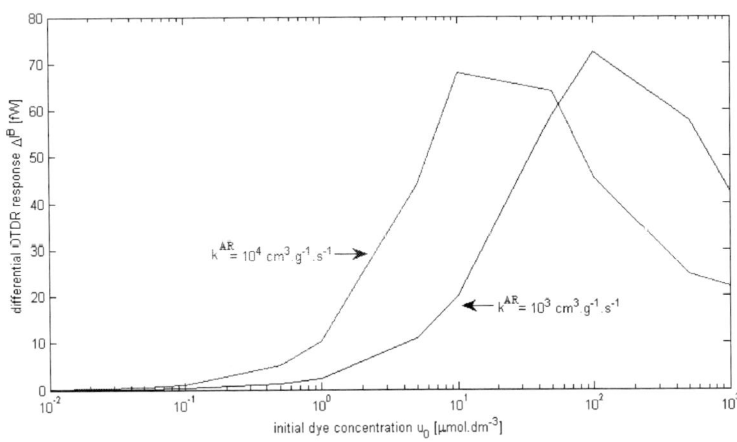

Figure 6. Dependence of the ammonia-induced $\Delta I^{B}(z)$ on the average reagent concentration within a 2 µm wide cladding zone, calculated for the two indicated reaction speed constants. Other parameters: $D = 10^{-9}$ cm^2/s, $v_0 = 5000$ ppm, exposition time 1000 s, middle of the doped zone 2 µm apart from the core/cladding boundary.

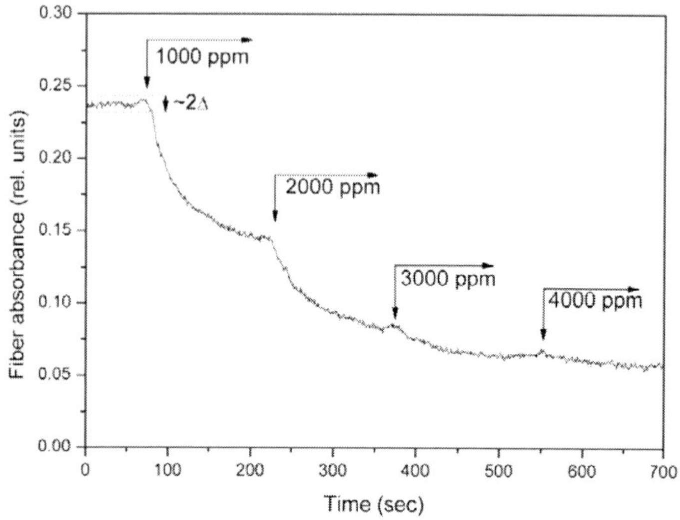

Figure 7. Change of optical absorption observed on short section of sensitized PCS optical fiber under consecutive exposition with the indicated concentration of ammonia gas in nitrogen. Reagent specification: Me = Co, A = Br and L = 5-(4'-dimethyl amino phenylimino) quinolin-8-one. Fiber specification: length 10 cm, b/a = 120 μm /100 μm. The signal recorded at λ = 730 nm using Ocean Optics S1000 spectrometer. The signal noise level is indicated limiting the initial sensor resolution to ~ 100 ppm of ammonia.

After a longer exposition period or in case of a high analyte concentration however, the reagent is 'depleted' and analyte molecules can approach the fiber core and stimulate strong sensor signal. An opposite situation can also occur. In case when the reagent concentration is low, molecules of reagent are quickly depleted by analyte and the sensor is loosing its performance. Example of such reaction is shown in Figure 7.

All the latter findings are very valuable from the point of view of a real optimized sensing fiber fabrication. However, a satisfactory knowledge and optimization of the reaction speed constant, especially its value for the case of reaction running in the cladding matrix seems to be an ultimate prerequisite. Practical preparation of a highly heterogeneous, spatially localized reagent distribution within a fiber cladding then constitutes another complicated challenge related to technology of optimized sensing fibers fabrication.

The developed model is also very instrumental in evaluation of real OTDR signals obtained from real sensing fibers. Figure 8 shows the OTDR curves recorded on a 120 m long multimode PCS optical fiber before and after its sensitization with organo-metallic reagent and then after exposition with ammonia gas.

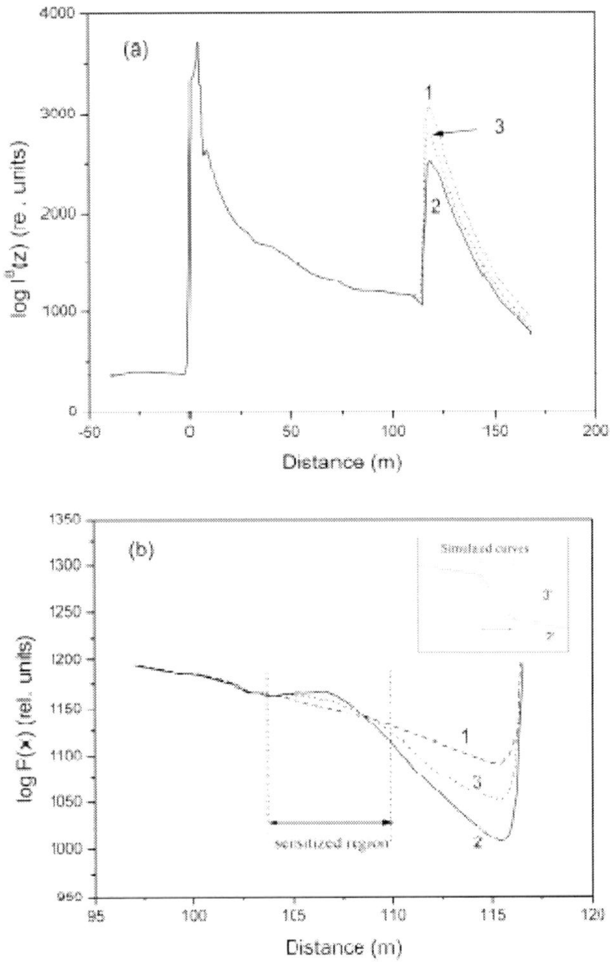

Figure 8. OTDR curves obtained with the 120 m long fibre sensitized within the distance x = 104 m -110 m; (a) - the overall courses, (b) – zoom of the sensitized region. Legend: curve 1 – the signal before sensitization; curve 2 – after sensitization with the reagent; curve 3 - after exposure to 10000 ppm ammonia gas in nitrogen; curves 2' and 3' - the simulated courses. Used reagent specification: Me = Cu, A = SO_4 and L = 5-(4'-dimethyl amino phenylimino) quinolin-8-one. Fiber specification: total length 120 m, b/a = 120 μm /100 μm. The OTDR signal recorded by Photodyne 5500 XFA (λ = 850 nm, τ = 20 ns) attached to HP 54615B digital oscilloscope.

For the testing purposes, the fiber was sensitized and exposed to ammonia/nitrogen mixture (10000 ppm of ammonia) within the interval of lengths 104 - 110 m. Impregnation of the fiber by the reagent was performed from its solution in ethanol/chloroform mixture. A non-optimized, linearly descending concentration profile following from the Fickean diffusion of reagent molecules into the polysiloxane cladding can be consequently expected to exist within the fiber cladding.

The observed OTDR signals show two types of well resolved reactions on both the sensitization and the exposition to ammonia. The first one relates to the intensity of the second Fresnel reflection from the fiber free end. After the sensitization, the reflection intensity is reduced as a result of the enhanced attenuation due to the reagent absorption. Exposition to ammonia, as expected, tends to reduce this effect (Figure 8a). It is worth to notice that examination of the second Fresnel reflection can be used as an instrumental tool for an 'integral' inspection of the distributed sensor.

The second reaction type relates to local changes of the OTDR course. In this case, the sensitization lead to increase of the signal within the sensitized region followed by a pronounced descent of the signal in comparison with the original state; again, ammonia exposition then results in a partial relaxation of this effects (Figure 8b). However, the observed courses significantly differ in shape from the simulated curves obtained using our model (cf. the inset in Figure 8b). In fact, the model only predicts the signal drop within and after the sensitized region.

The discrepancy very likely follows from several simplifications adopted in our model and demonstrate clearly restrictions of its application. To begin with, the current model form does not allow for simulation of the cladding RI increase caused by incorporation of the (highly polar) reagent molecules as well as for the RI reduction occurring during ammonia exposition. Such local increase of the cladding RI will cause a local reduction of the total mode number (as the result of *NA* reduction) accompanied by ascent of the light power transported within the cladding. The latter modal redistribution has again at least two different effects: (i) the local decrease in *NA* acts as a modal 'bottle-neck' and reduces the back-scattered intensity and (ii) the contribution to back-scattered intensity originating within the cladding volume growth (both due to the *NA* decrease and the N^R increase) resulting in a non-negligible observable contribution to the local scattered intensity. In fact, such effect is well known to relate to increase of cladding RI [48]. Quantification of the second effect is not possible with the current model for the exact formula describing the coefficient R_2 is missing.

In case of ammonia exposition, 'opposite' processes are taking place, but the situation is different from the initial, non-sensitized fiber: after the sensing reaction, the fiber cladding is still filled with the reaction products, ligand molecules and metallic ammonia complexes influencing on the fiber optical properties.

CONCLUSION

Application of the theoretical model developed for simulation of distributed FOS based on OTDR readout has proved to provide a mighty tool for design, optimization and functional analysis of such type of sensors. The obtained results confirm that development of the model and the experimental investigations of real systems must be performed in conjunction - the progress in one area is stimulating development in the second one.

The simulated results provide us with promise that one of the most serious drawbacks frequently discussed in relation to distributed fiber optic sensors, based on variations of optical absorption - the increase of transmission loss of the sensing fiber due to the sensitization with reagent - can be depressed using a carefully selected radial distribution of the reagent within the fiber cladding volume. New optical fiber sensitizing techniques capable of such precise structure fabrication must be developed in order to verify the model predictions.

Following further the simulations results, the speed constant of the employed sensing reaction appears to have a significant influence on the sensor function. Investigation of this issue and optimization of the reaction kinetics in the polymeric matrices used as cladding materials defines the next objective for future research.

Comparison of experimental data with the results of simulations has also revealed several weak points of the model. The most urgent corrections involve inclusion of effects pertinent to variations of the fiber cladding RI caused by chemical and physical processes accompanying the sensor operation, and formulation of the model extension correctly treating Rayleigh scattering contribution to the sensor signal rising up within the fiber cladding.

Application of the mentioned improvements and optimizations related to the distribution of reagent and its chemical kinetics has very likely the potential to move development of the distributed FOS of ammonia gas in the phase of a practical prototype fabrication and field testing.

ACKNOWLEDGEMENTS

Support of the presented research by grants MSM68407721 and MSM6840770040 of the Ministry of Education, Youth and Sports of the Czech Republic and grant SGS10/296/OHK4/3T/14 of the Czech Technical University in Prague is greatly acknowledged.

REFERENCES

[1] Timmer, B.; Olthuis, W.; van den Berg, A. *Sens. Actuat.* 2005, *B107, 666-677.*

[2] Koh, J.; Kim, B.; Hong, S.; Lim H,; Choi H. C. *J. Mater. Sci. Technol.* 2008, *24, 578-589.*

[3] Boshenkov V.E.; Sergeev G.B. *Russian Chemical Reviews* 2007, *76, 1011-1020.*

[4] Denualt, G. *Ocean Sci.* 2009, *5, 697-710.*

[5] Elia, A.; Di Franco, C.; Lugara, P.M.; Scamarcio, C. *Sensors* 2006, *6, 1411-1419.*

[6] Elia, A.; Lugara, P.M.; Di Franco, C.; Spagnob, V. *Sensors* 2009, *9, 9616-9628.*

[7] El-Sherif M.; Bansal L.; Yuan J. *Sensors* 2007, *7, 3100-3118.*

[8] Singh, M.; Verma, N.; Garg, A.K.; Redhu, N. *Sens. Act.* 2008, *B134 345–351.*

[9] Hibbard, T.; Killard, A. *Critical Reviews in Analytical Chemistry* 2011, *41, 21-35.*

[10] Jeronimo, P.C.A.; Araujo, A.N.; Conceicao, M.; Montenegro, B.S.M. *Talanta* 2007, *72, 13–27.*

[11] Orellana, G.; Haigh, D. *Current Analytical Chemistry* 2008, *4, 273-295.*

[12] R Moos, A. *Int. J. Appl. Ceram. Technol.* 2005, *2, 401–413.*

[13] Wang, S.G.; Zhang, Q.; Yang, D.J.; Sellin, P. J.; Zhong G.F. *Diamond Relat. Mater.* 2004, *13, 1327 -1332.*

[14] Zhang, T.; Nix, M.B.; Yoo, B.-Y.; Deshusses, M.A.; Myung, N.V. *Electroanalysis* 2006, *18, 1153-1158.*

[15] Qi P.; Vermesh O.; Grecu M.; Javey A.; Wang Q.; Dai H.; Peng S.; Cho K.J.; *Nano Lett.* 2003, *3, 347 -351.*

[16] Bochenkov, V.E.; Zagorski, V.V.; Sergeev G.B. *Sens. Act.* 2004, *B103, 375 -379.*

[17] Murray, B.J.; Li, Q.; Newberg, J.T.; Hemminger, J.C.; Penner, R.M. *Chem. Mater.* 2005, *17 6611 6618.*

[18] Raible, I.; Burghard, M.; Schlecht, U.; Yasuda, A.; Vossmeyer, T. *Sens. Act.* 2005, *B106, 730 -735.*

[19] Wang X.; Zhang, J.; Zhu Z. *Appl. Surf. Sci.* 2006, *252, 2404-2411.*

[20] Webber, M.E.; Hanson, R.K.; Jeffries, J.B. US Patent, *6,787,776.*

[21] A. Lobnik, A.; Wolfbeis, O.S. *Sens. Act.* 1998, *B51, 203-207.*

[22] Malins, C.; Doyle, A.; MacCraith, B.D.; Kvasnik, F.; Landl, M.; Simon, P.; Kalvoda, L.; Lukas, R.; Pufler, K.; Babusik, I. *J. Environ. Monit.* 1999, *1, 417-422.*

[23] Malins, C.; Butler, T.M.; MacCraith, B.D.; *Thin Solid Films* 2000, *368, 105-110.*

[24] Cao, W.; Duan, Y. *Sens. Act.* 2005, *B110, 252- 259.*

[25] Tao, S.Q.; Xu, L.; Fanguy, J.C. *Sens. Act.* 2006, *B115, 158-163.*

[26] Guo, H.; Tao, S. *Sens. Act.* 2007, *B123, 578-582.*

[27] He, Y.; Orr, B. *J. Appl. Phys. B: Lasers Opt.* 2006, *85, 355-364.*

[28] Galbarra, D.; Arregui, F. J.; Matias, I. R.; Claus, R. O. *Smart Mater. Struct.* 2005, *14, 739-744.*

[29] Kim, Y.-C.; Banerji, S.; Masson, J.-F.; Peng, W.; Booksh, K.S. *Analyst*, 2005, *130, 838-843.*

[30] Sharkany, J.P.; Korposh, S.O.; Batori-Tarci, Z.I.; Trikur, I.I.; Ramsden, J.J. *Sens. Act.* 2005 *B107, 77-81.*

[31] DuBois, S.; Eng, S.; Bhattacharya, R.; Rulyak, S.; Hubbard, T.; Putnam, D.; Kearney, D.J. *Digest. Dis. Sci.*, 2005, *50, 1780-1784.*

[32] Ritari, T.; Tuominen, J.; Ludvigsen, H.; Petersen, J. C.; Sorensen, T.; Hansen, T. P.; Simonsen, H. R. *Opt. Express*, 2004, *12, 4080- 4087.*

[33] Shahriary, M. R.; Zhou, Q.; Siegel Jr.G.H. *Opt. Lett.* 1988, *13, 407-409.*

[34] Blyler Jr.L.L. et al. *Polym. Eng. Sci.*, 1989, *29, 1215-1222.*

[35] Liebermann, R.A. *Proc. SPIE* 1991, *1586, 80-91.*

[36] Klein, R.; Voges, E. *Sens. Act.* 1993, *B11, 221-225.*

[37] Preininger, C. et al *Anal. Chim. Acta* 1996, *334, 113-123.*

[38] Grady T.; Butler T.; MacCraith B.D.; Diamond D.; McKervey M.A. *Analyst*, 1997, *122, 803-806.*

[39] Potyrailo, R.A.; Hieftje, G.M. *Anal. Chem.* 1998, *70, 3407-3412.*

[40] Nissilä, S. *Acta Universitatis Ouluensis C – Technica*, 1995, *83, 1-68.*

[41] Takada, K. *J. Lightwave Tech.* 2002, *20, 1001 – 1017.*

[42] Malins C.; Landl M.; Simon P.; MacCraith B.D. *Sens. Act.* 1998, *B51, 359 – 367.*

[43] Saleh B.E.A.; Teich M.C. *Fundamentals of photonics*; John Wiley and Sons: New York, US, 1991; pp 272-306.
[44] Ingle J.D.; Crouch S.R. Spectrochemical Analysis. Prentice Hall: New Jersey, US, 1988; pp 1 - 608.
[45] Sykora, O.; Aubrecht, J.; Klepacek, R.; Kalvoda, L. In *Computational Modeling and Advanced Simulations* Murin, J.; Kompis, V.; Kutis V.; Ed; Computational Methods in Applied Sciences; Springer Science and Business Media: New York, US, 2011; Vol. 24, pp 117-133.
[46] Chowdhury S.; Dodson J.; Bhethanabotla V.R. Sen R. In *Proceedings of AIChE Annual Meeting* 2007; ISBN 978-08169-1022-9.
[47] Lasaga A.C.; Cygan R.T. *Am. Miner.* 1982, *67, 328 – 334.*
[48] Bürck, J.; Sensfelder, E.; Ache, H.-J. *Proc. SPIE* 1996, *2836, 1-11.*

In: Ammonia: Structure, Biosynthesis... ISBN: 978-1-62100-502-5
Editors: V.A. Fekete, et al, pp. 61-89

Chapter 3

SPECIFIC INHIBITION BY AMINES AND AMMONIUM ION OF INITIATION AND ACTIVATION OF RIBOSOMAL RNA (rRNA) GENE EXPRESSION AT AND AFTER MIDBLASTULA TRANSITION (MBT) IN *XENOPUS* EMBRYOGENESIS

Koichiro Shiokawa*
Department of Judo Therapy, Faculty of Medical Technology;
Department of Biosciences, School of Science and Engineering,
Teikyo University; 1-1 Toyosatodai, Utsunomiya,
Tochigi Prefecture 320-8551, Japan.

ABSTRACT

In *Xenopus* embryogenesis, transcription of rRNA genes begins shortly after midblastula stage (or at the transition called MBT), and its activity increases greatly thereafter (Shiokawa et al., 1981a,b). We were interested in the control mechanism of this MBT-associated rRNA gene activation, and tried to find out substances which control rRNA gene

* Corresponding author: Koichiro Shiokawa; Department of Judo Therapy, Faculty of Medical Technology; Teikyo University; 1-1 Toyosatodai, Utsunomiya, Tochigi Prefecture 320-8551, Japan FAX, +81-28-627-7184; E-mail: shiokawa@nasu.bio.teikyo-u.ac.jp

expression. We examined the blastula cell conditioned medium, the blastula homogenate, and its acid-soluble fraction. After various tries and errors, we eventually reached the conclusion that weak bases inhibit quite selectively the synthesis of rRNA but not tRNA and mRNA and the inhibition takes place at the transcriptional level. In the present article, we first summarize our studies on initiation of rRNA gene expression during MBT, or the transition from the cleavage stage to the post-blastular stages in *Xenopus* developing embryos. We then summarize in some details our efforts performed to find out factors that control rRNA gene expression. Then, finally we describe our quite unexpected discovery that weak bases such as amines and ammonium ion selectively inhibit rRNA gene expression in *Xenopus* embryonic cells. We analyzed acid-extractable substances in *Xenopus* cleavage stage embryos and found that a larger amount of ammonia is present in pre-blastula stage embryos than in the post-blastular stage embryos. We also found that replacement of Na^+ with choline$^+$ in the culture medium completely abolishes the inhibition of rRNA gene expression. We, therefore, conclude that ammonium ion is one of the components that regulate rRNA gene expression in *Xenopus* embryogenesis, acting probably by inducing a slight increase in the intracellular pH.

Keywords: Xenopus embryogenesis, rRNA synthesis, 2'-O-methylation, mRNA CAP methylation, pre-MBT transcription, Specific inhibitor of rRNA synthesis, Weak bases, Amines, Ammonium ion.

1. Characteristic Features of *Xenopus* Embryos as an Experimental System

A *Xenopus* egg is a huge unique cell (1.2 mm in diameter) in which a number of cellular components are stored in the cytoplasm as maternal stockpiles for later use in development. Thus, a *Xenopus* egg contains large amounts of yolk granules and lipid droplets as nutrient sources, and other cytoplasmic organelles such as mitochondria (approx. 5×10^5/egg), ribosomes (approx. 10^{12}/egg), both being enough for at least 10^4 cells, and furthermore, contains RNA and DNA polymerases, histones, and maternal mRNAs, including those for caspases 9 and 3 (Takayama et al., 2004; Shiokawa et al., 2005), activin receptor (Kondo et al., 1991), SAMDC (S-adenosylmethionine decarboxylase) an enzyme for polyamine synthesis (Shinga et al., 1996), aldolases A and C for glycolysis (Kajita et al., 2000) and so on. Among the maternal stockpiles are also low-molecular-weight substances such as nucleotides and amino acids and ammonia (Shiokawa et al., 1986b).

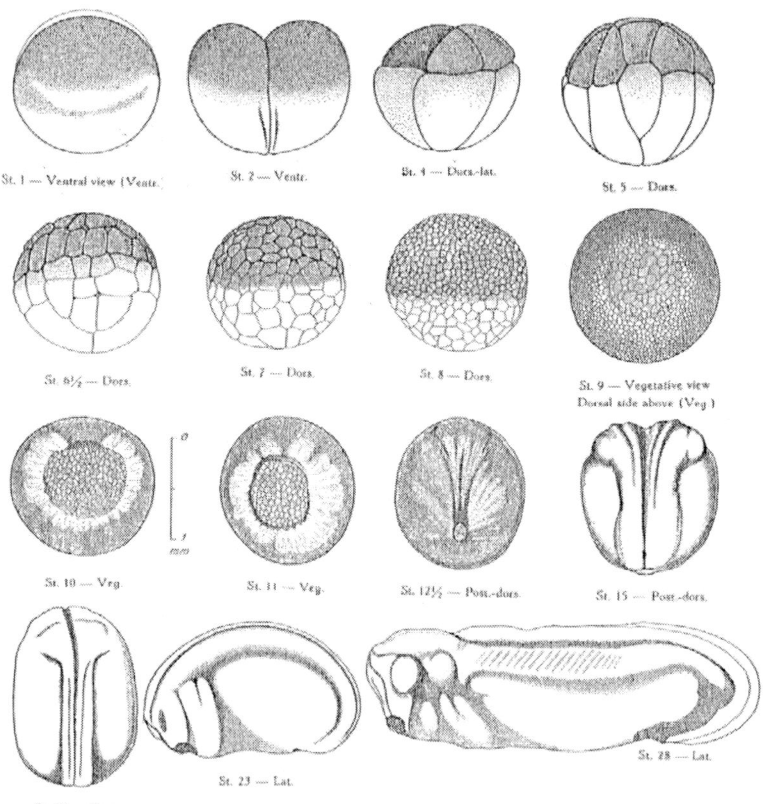

Pre-MBT: From St. 1 to St. 7 (Unfertilized egg stage to the end of cleavage stage).

Former half of MBT: From St. 8 to St. 8.5 (From early blastula stage to midblastula stage).

Latter half of MBT: From St. 8.5 to St. 9 (From midblastula stage to late blastula stage).

Post-MBT: St 10 to St. 28 (From late blastula stage to stage 28 late tailbud stage).

Figure 1. Early development of *Xenopus laevis*. St. 1 (Unfertilized egg); St. 2 (2-cell stage); St. 4 (8-cell stage); St. 5 (32-cell stage); St. 6.5 (Morula stage); St. 7 (Early blastula stage); St. 8 (Midblastula stage); St. 9 (Late blastula stage); St. 10 (Early gastrula stage); St. 11 (Midgastrula stage); St. 12.5 (Late gastrula stage); St. 15 (Early neurula stage); St. 20 (Neurula stage); St. 23 (Early tailbud stage); St. 28 (Late tailbud stage; Muscular response stage). From Nieuwkoop and Faber (1956).

In *Xenopus* embryogenesis, fertilized eggs cleave rapidly and synchronously without G_1 and G_2 phases in the cell cycle, and this type of rapid and synchronous cell divisions continue for at least 12 rounds, then embryos reach the blastula stage (Figure 1). At the blastula stage, however, cell division become asynchronous, with the appearance of both G_1 and G_2 phases (Newport and Kirschner, 1982). During the early phase of the development, translation of maternal mRNA, which has been accumulated during oogenesis, takes place actively (Richter et al., 1982). Transcription from nuclear genes, however, is not easily detected during the cleavage stage, although it does occur to a substantial extent when the nuclear transcriptional activity was expressed on a per-cell basis, as opposed to per-embryo basis (Shiokawa et al., 1981a; b; 1994; Brown and Littna, 1966; Shiokawa and Yamana, 1965; 1967a). The main reason for the detection of only a negligible low level per embryo of RNA synthetic activity is the presence within an embryo of a very small number of nuclei (or cells) especially in embryos at early cleavage stages. Another reason for the difficulty to detect new transcription in early *Xenopus* embryos is that, unlike embryos of other animal species such as sea urchin, *Xenopus* embryos have impermeable surface coat on the cytoplasmic membrane, which prevents uptake of various radioactive RNA precursors added into their culture medium (Shiokawa and Yamana, 1967; Shiokawa et al., 1967).

After the 12 rounds of cleavage cell cycles, embryos reach midblastula stage (4096 cells/embryo), or the stage of midblastula transition (MBT)(Newport and Kirschner, 1982), after which G_1 and G_2 phases reappear in the cell cycle (Heasman, 2006).

At the MBT, various other changes in cellular activities take place (Shiokawa et al., 1989; Newport and Kirschner, 1982; Yasuda and Schwiger, 1992; Andeol, 1994; Yang et al., 2002), including initiation of checkpoint-regulated cell cycle control (Wroble and Sible, 2005; Carter and Sible, 2003), acqusition of cell motility (Newport and Kirschner, 1982; Minoura et al., 1995), and increase in zygotic nuclear transcriptional activity (Shiokawa et al., 1994; 1989; Yasuda and Schubiger, 1992; Yang et al., 2002; Nakakura et al., 1987; Shiokawa et al., 1981a,b).

Besides the characteristic features described above, *Xenopus* early embryos have another characteristic feature in their development. It is the maternal program of apoptosis preset in early embryos. We found it when we cloned *Xenopus* homologue of SAMDC and tested the function of SAMDC cDNA by microinjecting its *in vitro*-synthesized mRNA (1 ng/egg) into fertilized egg cytoplasm. Since the level of endogenous SAMDC mRNA in

Xenopus is 0.005 ng/embryo or less, the microinjection of 1 ng of SAMDC mRNA results in the overexpression of the mRNA at more than 200-folds, and the microinjection induced a corresponding overexpression of SAMDC enzyme activity.

The SAMDC-overexpressed embryos cleaved normally up to the early blastula stage, but at MBT, they were suddenly dissociated into cells and completely dissolved (Shibata et al. 1998). SAMDC mRNA was effective at 0.1-100 ng/egg, and in this wide range of dosage, cell dissociation took place always at MBT, suggesting the occurrence of some clock mechanism for determination of the timing of the cell dissociation. In the SAMDC-overexpressed embryos, protein synthesis was the first to be inhibited and later DNA and RNA synthesis were also inhibited. When we injected SAMDC mRNA together with ethylglyoxal-bis(guanylhydrazone) (EGBG), a specific inhibitor of SAMDC, the cell dissociation was completely suppressed and embryos developed into swimming tadpoles.

In such embryos the level of SAM (S-adenosylmethionine) was extremely low and when SAM (200 pmoles/egg) was injected together with SAMDC mRNA, cell dissociation was completely suppressed and embryos developed into tadpoles. Electron microscopic analyses revealed that nuclei of such dissociated cells were fragmented, and in such embryos a large number of cells became TUNEL-positive, and DNA extracted therefrom formed "ladders" (Kai et al. 2000; Kaito et al. 2001), and furthermore the cell-dissociating effect was suppressed by prior microinjection of Bcl-2 mRNA. We, therefore, concluded that the SAMDC-induced cell dissociation was due to the execution of apoptosis (Kai et al. 2000; Shiokawa et al. 2000). In the SAMDC-overexpressed embryos, we analyzed caspases, a cysteine protease enzyme family, and found that caspases 8 and 9, and also 3 are involved in this order in the execution of the apoptosis (Takayama et al. 2004; Shiokawa et al. 2005; 2008). mRNAs for caspases 9 and 3 occurred as maternal mRNAs, whereas caspase 8 mRNA was found to be newly transcribed in the SAMDC-overexpressed embryos before MBT (Shiokawa et al. 2005).

From the experiments to follow the fate of a blastomere that received GFP and SAMDC mRNAs at 8-32 cell stages, we reached a conclusion that the apoptosis in *Xenopus* early embryos serves as a fail-safe mechanism of early embryogenesis to eliminate metabolically-damaged cells at MBT to save the rest of the embryo in the subsequent development (Kai et al. 2003; Shiokawa et al. 2008) (Figure 2).

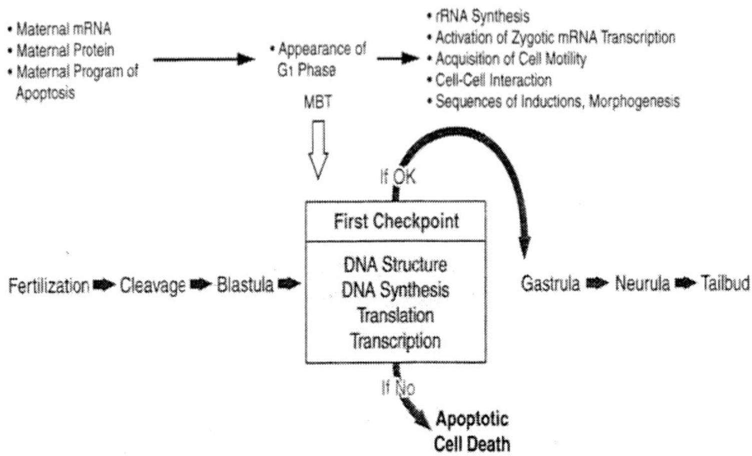

Figure 2. Various changes which take place during *Xenopus* early embryonic development, with special reference to the maternally preset program of apoptosis. This model shows how early development proceeds. This model indicates the occurrence of apoptotic checkpoint at the step of the midblastula stage or the stage of midblastula transition (MBT) which functions as a surveillance or "fail-safe" mechanism in *Xenopus* early embryonic development. Fertilized eggs cleave rapidly until the early blastula stage. At MBT, the "first developmental checkpoint" comes when G_1 phase first appears. We assume that this check mechanism determines cell-autonomously if the cell continues or stops development to be eliminated by execution of the maternally inherited program of apoptosis. However, even when apoptosis was executed, embryos follow two different courses. If the number of apoptotic cells was large, the whole embryo stops development and dies. However, if the number of apoptotic cells was small, they are confined within the blastocoel and the embryo itself continues on development. Cells to be eliminated are segregated into the blastocoel. If such cells came out to the perivitelline space, the whole embryo will be dissolved due to osmotic shock (eggs are laid in the water), and the maternally inherited program of apoptosis will not serve as a fail-safe mechanism. From Shiokawa et al. (2000).

2. Dissociated *Xenopus* Embryonic Cells as a New Experimental System Suitable for Studying rRNA Gene Expression

Ribosomes are supramolecular structures which are large cellular machinery to produce proteins. Cellular activity to produce ribosomes sensitively reflects the cell physiological conditions for cell growth (Lieberman et al., 1963). In eukaryotic cells ribosomes consist of 60S and 40S particles and RNA moiety of 60S particles consists of 28S rRNA and 5.8S RNA and 5S RNA, and that of 40S particles consists only of 18S rRNA. In *Xenopus* oocytes rRNA producing genes (rDNAs) are specifically amplified at stage III of oogenesis (Brown and Dawid, 1968), and mature oocytes accumulate ca. 10^{12} ribosomes which are enough for supporting protein synthesis for development up to the stage 42 tadpoles (Gurdon and Brwon, 1965).

Brown first studied ribosomes and ribosomal RNA (rRNA) in *Rana pipiens* embryos using a sodium acetate-NaOH (pH 5.0) as an RNA extraction buffer (Brown and Caston, 1962), but he soon changed the material from *Rana pipiens* to *Xenopus laevis* (Brown and Littna, 1964). Then, Gurdon and Brown (1965) studied rRNA synthesis in embryos of anucleolate mutants of *Xenopus laevis.* They showed that the mutant embryos neither formed nucleoli nor synthesized rRNA, and thus proved that nucleoli formation is the cytological manifestation of rDNA function. In Japan, we used *Rana japonica* embryos for study of RNA synthesis and isolated undegraded, as opposed to alkali-hydrolyzed, embryonic RNA using a Tris-HCl RNA extraction buffer (pH 7.2) (Shiokawa and Yamana, 1965). According to Brown and Littna (1964), we also changed the material from *Rana* to *Xenopus*, and using the sodium acetate buffer (pH 5.0) as an RNA extraction buffer, started to study *Xenopus* embryonic RNA (Shiokawa and Yamana, 1965; 1967; Shiokawa et al., 1967). In our studies, we used vitelline membrane-depleted embryos or dissociated embryonic cells, instead of whole embryos, as a new experimental system suited for the studies of *Xenopus* embryonic RNA synthesis (Figure 3). Frog embryos have impermeable surface coat, and usually they can not be labeled with radioactive precursors added in their surrounding medium (Brown and Littna, 1964). However, vitelline membrane-depleted embryos and dissociated embryonic cells were easily labeled by simply adding radioactive precursors of RNA (either ^{3}H-uridine or ^{3}H-adenosine), DNA and protein into their culture medium, and this was a much easier method to label newly-synthesized RNA as compared with the previous method to use $^{14}CO_2$ in a closed vial (Cohen, 1954). Shiokawa and Yamana (1967a) compared the overall RNA synthetic patterns in ^{3}H-uridine-labled dissociated cells and $^{14}CO_2$-labeled whole embryos, and reached the conclusion that developmental changes in the pattern of rRNA synthesis is the same between the dissociated cells and whole embryos. Thus, dissociated blastula cells synthsize RNA just as blastulae do, and the blastula cells change their RNA synthetic pattern from the blastula-type to the neurula-type just as blastulae do when they become neurulae (Shiokawa and Yamana, 1967a).

Figure 3. Embryonic cells used for isotopic labeling experiments. A 64-cell stage embryo from which vitelline membrane was removed (top) and cells dissociated from embryos at the blastula stage (bottom). In the top figure, A and V are for animal side cells and vegetal side cells, respectively. (A) is by the courtesy of Dr. Kosuke Tashiro, Kyushu University. (B) is from Shiokawa and Yamana (1967).

3. A Unique Developmental Regulation of rRNA Gene Expression before and after Midblastula Stage, or the Stage of MBT

Ribosomal RNA synthesis was first considered to be initiated at the gastrula stage during *Xenopus* embryogenesis (Gurdon, 1974). However, evidence that rRNA synthesis does not occur in pre-gastrular stages was not completely conclusive, because RNA fractionation methods routinely used in previous experiments were sucrose density gradient centrifugation and agarose-polyacrylamide gel electrophoresis, and these do not completely separate rRNA from heterogeneous mRNA-like RNA, which happens to be synthesized very actively in pre-gastrular embryos (Shiokawa et al., 1981a,b). 18S and 28S rRNA molecules contain a relatively large number of 2'-O-methyl groups (ca. 90 per 40S pre-rRNA, almost all of which is conserved in the processed rRNAs), whereas most eukaryotic mRNAs contain no 2'-O-methyl group except in the cap terminus position. Therefore, rRNA and mRNA can be separatedly quantitated if 2'-O-mrthylated nucleotide groups were determined in nuclease digests of these RNAs. Since rRNA precursor is methylated immediately after its transcription, the procedure to detect rRNA-specific 2'-O-methylation detects rRNA gene product as soon as it is transcribed. This means that the presence of contaminating mRNA does not affect rRNA determination, and separation of rRNA from mRNA is not necessarily an essential step of the fractionation in this particular case. According to this principle, we labeled *Xenopus* embryos with (methyl-^{3}H)methionine at the morula, blastula, gastrula and neurula stages, purified their ^{3}H-labeled high-molecular-weight RNAs by sucrose density gradient centrifugation, and analyzed their nucleotide digests on DEAE-Sephadex columns. From these experiments, we found that embryos start to synthesize rRNA during the 4 hr of blastula stage. We then further divided this 4 hr-period into two 2 hr-periods, and again studied rRNA-specific 2'-O-methylation after labeling embryos with (methyl-^{3}H)methionine in each of the two 2 hr-periods. The results obtained showed that embryos contained only a trace amount of rRNA-specific dinucleotides in the first 2 hr of the blastula stage but contained a substantially large amount of rRNA-specific dinucleotides in the latter 2 hr, indicating that rRNA synthesis starts in the latter half of the MBT stage (Figure 4). By contrast, the extent of cap formation and base-methylation were not greatly different in the former and the latter half periods of blastula stage, indicating that rRNA synthesis is activated only during the latter 2 hr period of the blastula stage. From these results we concluded that rRNA synthesis starts during the mid-to late blastula stage (Shiokawa et al., 1981a,b). The onset of rRNA gene expression in the late blastula stage is consistent with the first appearance of nucleoli at this stage (Nakahashi and Tamana, 1976), and also consistent with the timing of the first detection of the transcription of the exogenously-microinjected rDNA in *Xenopus* embryogenesis (Busby and Reeder, 1983).

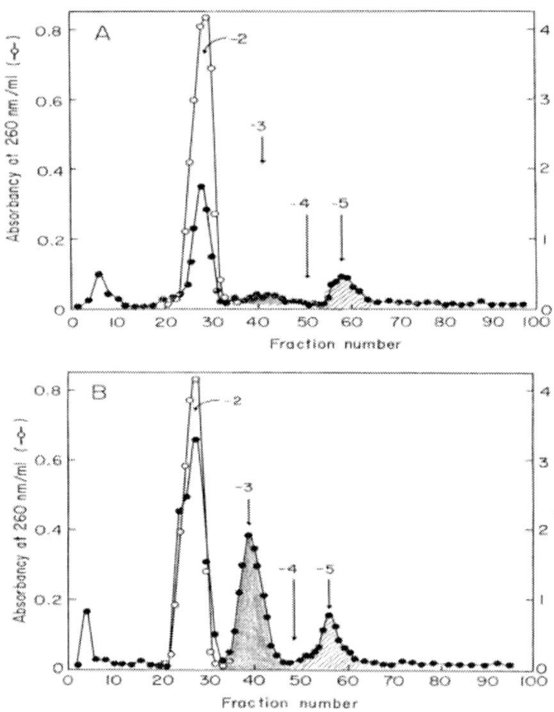

Figure 4. Comparison of RNA methlation patterns between the former half and the latter half of blastula stage. (A) Blastula cells were labeled for 2 hrs with (methyl-^{3}H)methionine immediately after cell dissociation. (B) Blastula cells were first cultured for 2 hrs, then labeled as in (A) for 2 hrs. RNA was extracted and DEAE-Sephadex column chromatography was performed as to find out 2'-O-methylatiom. Charge value -3 component which is specific for rRNA synthesis, or rRNA-specific dinucleotides ($N^{m}{}_{p}N_{p}$) , occurs only in the latter half of the blastula stage, whereas the labeling of charge value -5 substance, which is mRNA cap-specific methylation, or methylated type I cap structure ($m7G_{ppp}N^{m}{}_{p}N_{p}$) occurred equally in both (A) and (B). The -4 charge value component is for methylated trinucleotides ($N^{m}{}_{p}N^{m}{}_{p}N_{p}$). From Shiokawa et al (1981a).

From the rates of incorporation of the (methyl-3H)-group into rRNA specific dinucleotides ($N^{m}{}_{p}N_{p}$) and mRNA-specific cap structure, and the specific radioactivity of S-adenosylmnethionine which was determined by separate column chromatographic analysis, the rates of syntheses of 18S rRNA and 28S rRNA and capped mRNA were estimated. The rate of rRNA synthesis per embryo was about 1 ng/embryo/hr at the blastula stage when rRNA synthesis starts, and this rate per embryo continued to increase greatly throughout later stages. The rate per cell, on the other hand, increased only during the first 2 hr after onset of rRNA synthesis from an initial level of 0.02 pg/cell/hr to about 0.12 pg/cell/hr, and the there was little further increase in the subsequent 15 hrs of development.

4. Timing of the Initiation of mRNA Cap Methylation Is Not the Blastula Stage but the Cleavage Stage in *Xenopus* Embryogenesis

It is apparent from the data in Figure 4 that capped mRNA synthesis starts in the pre-blastula stage (detailed data is not shown, but actually from 32 cell stage through morula to early blastula stage) unlike rRNA synthesis. We confirmed the absence of the labeling of rRNA-specific 2'-O-methylation in morula stage by laborious labeling of cells from as many as 3,000 cleavage embryos to show that rRNA is not synthesized at the morula stage. In this experiment, the failure to detect rRNA-specific di- or tri-nucleotides in the early blastula stage was not due to the fact that early blastula stage embryos contain a much smaller number of cells than embryos in the late blastula stage embryos, since in this labeling experiment the number of embryos used at different stages were adjusted so that the number of labeled cells, as opposed to the number of labeled embryos, became approximately the same (in the order of 10^6) throughout stages. Synthesis of heterogeneous mRNA-like RNA has been suggested to occur in embryos at the cleavage stage by Brown and Littna (1964), and as they suggested, we, for the first time in 1981, detected the mRNA like RNA synthesis in morula stage by the methods to detect mRNA cap methylation (Shiokawa et al., 1981a,b). Also, this data was confirmed later in 1987 by the first detection by ^{3}H-uridine-labeling of heterogeneous mRNA-like RNA in the morula stage (Nakakura et al., 1987).

5. Attempts to Find out a Factor That Controls rRNA Gene Expression in *Xenopus* Embryonic Development

As for the mechanism of the developmental control of rRNA gene expression, the classical nuclear transplantation experiment by Gurdon and Brown (1965) strongly suggested that a cytoplasmic factor is involved, because a nucleus from later stage embryos transplanted into the unfertilized egg stopped synthesizing rRNA during cleavage, and nucleoli within the nucleus disappeared until they are restored later at around the gastrula stage. Shiokawa and Yamana (1967b) prepared dissociated blastula cells and neurula cells simultaneously, and after culturing them together or separately, obtained data which suggested that *Xenopus* blastula cells may release some factor that inhibits rRNA synthesis in neurula cells, by examining the incorporation of exogenously-added ^{3}H-uridine into rRNA in neurula cells. Since untreated blastula cell-conditioned medium and its heat-treated sampe appeared to be equally inhibitory on neurula cell rRNA synthesis, we assumed that the blastula-conditioned medium contains a low-molecular-weight substance that inhibits rRNA synthesis in neurula cells. In these experiments, RNA analyses were performed in most cases using MAK (methylated albumin-kieselguhr) columns and, in this system, while 18S and 28S rRNAs were eluted together as a single peak from the column by 0.6-0.8 M NaCl gradient, 4S RNA (most of which was tRNA) was eluted from the column as a stepwise fraction using 0.4 M NaCl. Therefore, in the initial step of the experiments, while the estimation of rRNA labeling was performed relatively accurately, that of 4S RNA might have not been accurate enough because of the inclusion of substances which are smaller than 4S RNA, thereby introducing overestimation to a certain extent of 4S RNA synthesis in the conditioned medium-treated embryos. Anyway, the experiments performed utilizing the conditioned medium were not of high reproducibility (Shiokawa and Yamana, 1975b; Laskey et al. 1973).

Therefore, we made blastula homogenate, and we started to use 0.5 N perchloric acid (PCA)-extractable materials as a starting materials for purification of the putative inhibitor of rRNA synthesis in *Xenopus* blastulae or cleavage stage embryos. Thus, we homogenized early embryos with 0.5 N PCA and after neutralizing it with KOH, applied it onto a small charcoal column and eluted the colum with ammonia-alcohol (0.2 N ammonia-50% ethanol). The white powder obtained by evaporation of the ammonia-alcohol eluate from the charcoal column was found to reproducibly and selectively inhibit rRNA synthesis in neurula cells (Shiokawa and Yamana, 1975b). Since we suggested this methods by personal communication, Laskey et al. (1973) obtained the same results and their results were pulished in Gurdon's textbook (Gurdon, 1974) (Figure 5).

In our efforts to purify and identify the active substance from the charcoal extract, we could not succeeded in obtaining a natural substance that inhibits rRNA synthesis selectively. However, we finally discovered that the active substance in the charcoal eluate was ammonium perchlorate that was artifactually formed during the charcoal column chromatography. Figure 6 is a set of the data which shows that ammonium perchlorate purified from the charcoal extract strongly and selectively inhibits rRNA synthesis in *Xenopus* neurula cells (Shiokawa et al., 1985).

This was a rather discouraging finding. However, the selective inhibition of rRNA synthesis induced here by the ammonium perchlorate was quite clear and interesting: the synthesis of 4S RNA occurred almost normally in spite of the strong inhibition (more than 70%-80%) of the synthesis of rRNA. Then, we further continued our studies on this line, concentrating on the effects of weak bases, in general, on rRNA synthesis in *Xenopus* embryonic cell system.

6. Weak Bases (Ammonium Ion and Amines) Selectively Inhibit rRNA Gene Expression in *Xenopus* Neurula Cells

The concentration of ammonium perchlorate, which was selectively inhibitory on rRNA synthesis, turned out to be ca. 2.5 mM. Therefore, we first tested the effects of the same concentration of sodium perchlorate on rRNA synthesis in *Xenopus* neurula cells, with the finding that this was totally ineffective as a selective inhibitor of rRNA synthesis. This implied that the effective part of the ammonium perchlorate is not the perchlorate moiety, but the ammonium moiety.

We then tested effects of various inorganic ammonium salts on RNA synthesis in Xenopus neurula cells. We cultured Xenopus neurula cells in the medium containing 2.5 mM of ammonium chloride or ammonium sulfate, and found that both of these inhibit rRNA synthesis, with little inhibition on 4S RNA synthesis (Shiokawa et al., 1985; 1986a,b). Also, ammonium dihydrogenphosphate, ammonium monohydrogenphosphate, and ammonium phosphate were all effective at 2.5 mM as selective inhibitors of rRNA synthesis in neurula cells. The effectiveness appeared here to be roughly proportional to the number of ammonium ions to be released. We also tested the effect of potassium dihydrogenphosphate, potassium monohydrogenphosphate and potassium phosphate, but again none of these were effective. Organic ammonium salts such as ammonium acetate and ammonium aspartate were all effective at 2.5 mM as selective inhibitors of rRNA synthesis. By the charcoal column chromatography, various nucleotides were concentrated at the step of the chromatography, and we repeatedly confirmed that the ammonia-alcohol eluate containing such nucleotides were always effective in selectively inhibiting rRNA synthesis. Therefore, we tested ammonium salts of all the major ribonuleotide monophosphates (5'-AMP, 5'-GMP, 5'UMP, and 5'-CMP) at 5 mM, and found that these were also effective in selectively inhibiting rRNA synthesis in neurula cells.

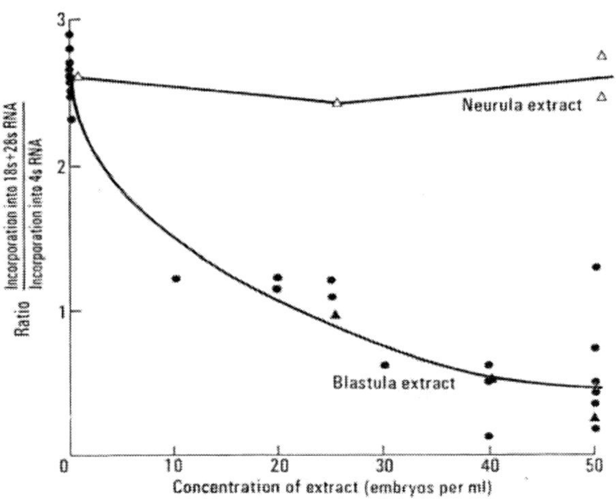

Figure 5. Results of extensive analysis performed by Laskey, Kirschner, and Knowland (1973) for a cytoplasmic inhibitor of rRNA synthesis. Eggs or embryos of *Xenopus laevis* were extracted with perchloric acid. The extract was partially purified by charcoal adsorption (This charcoal purification was the original methods by Shiokawa and Yamana but since Laskey et al. did not succeed in detecting a selective inhibitor of rRNA synthesis in blastula conditioned medium, Shiokawa and Yamana suggested to use this charcoal methods to get a concentrated inhibitor preparation from preblastula stage embryos) and then added to isolated cells of neurula embryos. If the extract was prepared from blastulae which do not synthesize rRNA, it inhibited rRNA synthesis in neurula cells. If the extract was prepared from neurulae which normally synthsize rRNA, it failed to inhibit rRNA synthesis. From Gurdon (1974).

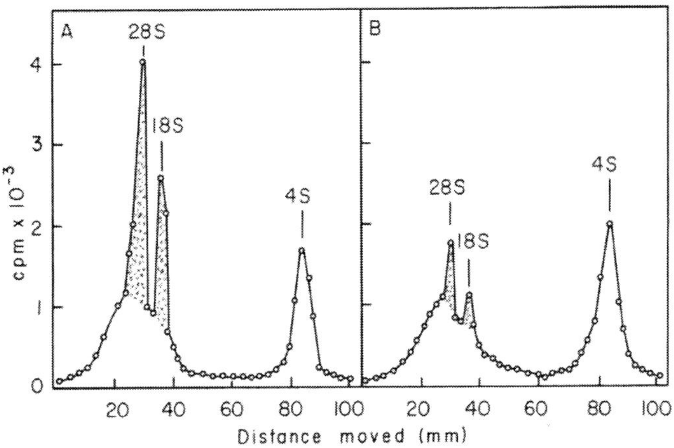

Figure 6. Effects of a chloroform-methanol extract of a charcoal fraction on RNA synthetic pattern of neurula cells. Cleavage stage embryos were homogenized with 0.5 N perchloric acid, and the acid-soluble fraction obtained was neutralized with 0.5 N KOH. The charcoal column was applied with the neutralized acid-soluble faraction, washed with water, and then eluted with ammonia-alcohol. The eluate was dried by lyophilization, and while powder finally obtained was extracted with chloroform-methanol. Neurula cells were labeled with (^{3}H)uridine in the presence or absence of the chloroform-methanol extract. RNAs were extracted and electrophoresed on agarose-polyacrylamide gels. (A), Control neurula cells. B, Treated neurula cells. Peaks of 18S and 28S rRNAs are dotted. From Shiokawa et al. (1985).

We further tested effects of 2.5 mM of primary, secondary and tertiary methylamines neutralized by hydrochloric acid or hydroperchloric acid. We found here that monomethylamine hydroperchloride, dimethylamine hydroperchloride, and trimethylamine hydrochloride were all active as selective inhibitors of rRNA synthesis. Similar positive results were obtained with three kinds of ethylamines. We tested polyamines such as putrescine, spermidine and spermine, but nine of these were effective as a selective inhibitor of rRNA synthesis even at 10 mM. Alpha-methylornithine that inhibits ornithine decarboxylase (Emanuelsson and Heby, 1978) has been claimed to be a direct stimulator of rRNA synthesis (Russell, 1983). When neurula cells were incubated in the presence of alpha-methylornithine, the amount of ornithine increased by about 3 times (61 ng per embryo) that of the control embryo (22 ng/embryo). Nevertheless, we found neither stimulative nor inhibitory effect on rRNA synthesis.

Using amino acid analyzer, we quantitatively analyzed ammonia and amines in acid-soluble fractions of *Xenopus* neurula cells which had been treated with 2.5 mM of either ammonium chloride, three kinds of ammonium phosphates, or three kinds of methylamines. In the amino acid profiles, it has been shown that the peak representing ammonia (peak "o" in Figure 7) became quite large in the cells treated with ammonium salts. In cells treated with monomethylamine hydroperchloride, a large amount of monomethylamine (peak 'r' in Figure 7C) was detected, with no increase in the amount of ammonia (peak "o"). However, amounts of other components such as ornithine and other amino acids in the neurula cells did not change appreciably in spite of the treatments with ammonium salts and amines. These results indicate that weak bases like ammonium ion and amines which selectively inhibit rRNA synthesis are uptaken into *Xenopus* neurula cells.

7. Weak Bases Suppress Both Initiation and Activation of rRNA Gene Expression Which Starts at and after MBT Stage in *Xenopus* Embryogenesis

As described above, weak bases inhibit rRNA gene expression in the neurula cells, in which rRNA synthesis is fully activated. The next question we asked was if weak bases suppress the activation of rRNA gene expression at MBT in *Xenopus* embryogenesis. We cultured blastula cells for 5, 10, or 15 hrs in the medium that contained 2.5 mM of ammonium salts, and during the culture we labeled the blastula cells with ^{3}H-uridine for 5 hrs at different timing. When control blastula cells were labeled for the first 5 hrs, only a small amount of incorporation was detected in 28S and 18S rRNA with a large amount of incorporation into 4S RNA and DNA, reflecting that rRNA synthesis at this stage is only at the step of the initiation of the expression (Figure 8A,B,C).

It is shown here that while ammonium chloride inhibited the activation of the rRNA synthesis, potassium chloride did not. Much clearer inhibition of rRNA synthesis was observed when blastula cells were first cultured for 5 hrs in the presence of 2.5 mM of ammonium chloride and then labeled for the next 5 hrs in the continued presence of the ammonium chloride (Figure 8D,E,F).

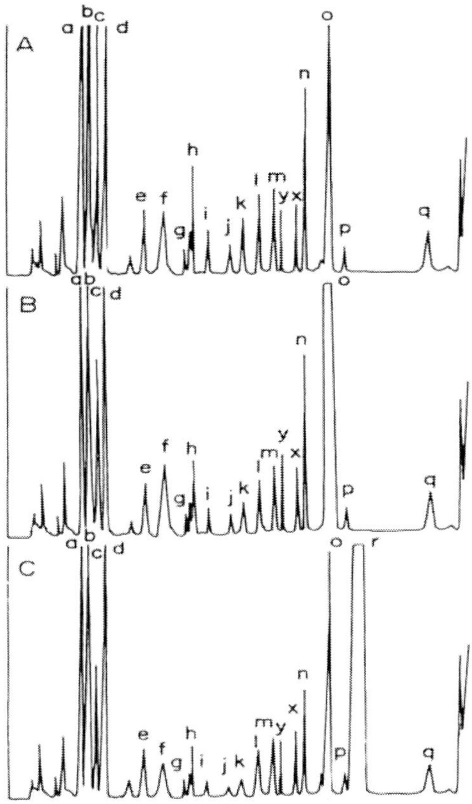

Figure 7. Analysis on amino acid analyzer of acid-soluble fractions of neurula cells treated with ammonium salts and amines. Cells from 200 neurulae were treated with 2.5 mM of ammonium dihydrogenphosphate, monomethylamine hydroperchloride for 5 hr. (a) asparagine plus aspartic acid, (b) threinine, (c) serine, (d) glutamine, (e) glycine, (f) alanine, (g) cysteine, (h) valine, (i) methionine, (j) isoleucine, (k) leucine, (l) tyrosine, (m) phenylalanine, (n) lysine, (o) ammonia, (p) histidine, (q) arginine, (r) monomethylamine, (x) ornithine, (y) γ-aminobutylic acid. (A) Control, (B) Treated with ammonium dihydrogenphosphate, (C) Treated with monomethylamine hydroperchloride. From Shiokawa et al. (1986).

Essentially the same results were obtained also when blastula cells were treated with ammonium chloride for 10 hrs and then labeled for 5 hrs in the continued presence of ammonium chloride (Figure 8G,H,I), although in this last experiment there was some inhibition in the cellular activity to form cellular reaggregates. Throughout the experiment syntheses of both 4S RNA (tRNA) and heterogeneous nuclear RNA were again not interfered greatly.

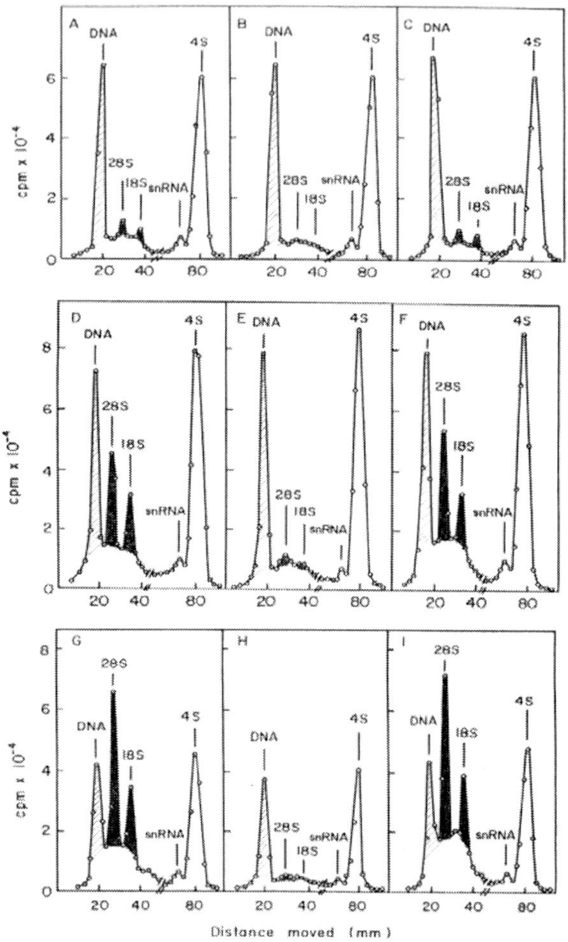

Figure 8. Inhibition of the initiation and activation of rRNA gene expression by ammonium chloride but not by potassium chloride in *Xenopus* blastula cells. Blastula cells were treated with 2.5 mM of either ammonium chloride (B, E, H) or potassium chloride (C, F, I) and labeled with 3H-guanosine. (A-C) Cells were labeled for 5 hr immediately after commencement of the treatment. (D-F) Cells were first cultured for 5 hrs in the presence of ammonium chloride or potassium chloride and then labeled with ^{3}H-guanosine in the continued presence of the salts. (G-I) Cells were first cultured for 10 hrs in the presence of the salts and then labeled for 5 hrs in the continued presence of the salts. RNAs were extracted and electrophoresed on 0.5% agarose-2.4% polyacrylamide gels. In this profile the radioactivity appeared also in DNA, since the tracer used was 3H-guanosine which can be incorporated into both DNA and RNA. (A, D, G) Control cells. From Shiokawa et al. (1986).

Using 10% polyacrylamide gel, we further discriminated small-molecular-weight RNAs in the same RNA preparations and confirmed that ammonium chloride did not inhibit the labeling of not only 4S RNA but also 5S RNA, and U_1, U_2, and U_5 small nuclear RNAs. We also confirmed that when ammonium ion was eliminated from the medium and the culture was incubated further in the absence of ammonium ion, rRNA synthesis can be restored to a large extent, indicating that the effect of ammninium salts is reversible. We also confirmed that ammonia given in the culture medium of neurula cells as ammonium chloride was incorporated into neurula cells within 1 hr after the treatment. We also confirmed that ammonium salts did not inhibit DNA synthesis, protein synthesis, cell division, and cellular reaggregating activity at least for the first 10 hrs of incubation. These data clearly show that ammonium salt inhibits activation at MBT and subsequent increase in the activity of rRNA synthesis.

8. Weak Bases Inhibit rRNA Gene Expression at the Transcription Level in *Xenopus* Embryonic Cells

To know the mechanism by which weak bases inhibit rRNA synthesis, *Xenopus* neurula cells were first treated for 2.5 hrs by 5 mM of potassium chloride (as a negative control), ammonium chloride or monomethylamine hydroperchloride, and then pulse-labeled for 2.5 hrs in the continued presence of these weak bases. In the control cells we obtained active labeling of heterogeneous nuclear RNA (hnRNA), 40 S rRNA primary transcript, and 4S RNA, in addition to the very small amount of labeling of 28S and 18S mature rRNAs in the control cells, whereas in the treated cells both of the weak bases inhibited the labeling of only 40S pre-rRNA, indicating that the inhibition is at the transcription level.

Under the conditions of partial inhibition of rRNA synthesis, we tested the effects of ammonium chloride and trimethylammonium perchloride on rRNA processing. When we labeled the control untreated neurula cells for 1 hr, we obtained the labeling of 40 S pre-rRNA, 30 S rRNA intermediate and 18S matured rRNA, in addition to a large amount of heterogeneous mRNA-like RNA. When neurula cells treated with 2.6 mM ammonium chloride for 2.5 hrs were labeled with 3H-uridine for 1 hr, the labeling of 40s pre-rRNA, 30s rRNA intermediates and 18S rRNA was inhibited by 60%, 59%, and 70%, respectively, as compared with the control. When the same cultures were labeled for 3 hrs, the inhibition of the labeling of 18 S and 28 S mature rRNA, and 4S RNA was 72%, 69%, and 7%, respectively. We obtained the similar results in a parallel experiment using 1 mM trimethylammonium perchloride as an inhibitor: the inhibition was 67%, 70%, 75%, 73% and 8%, for 40S pre-rRNA, 30S rRNA intermediate, 28S rRNA, 18S rRNA and 4S RNA, respectively. Thus, the inhibition observed in primary rRNA transcript (40S RNA) and its processing intermediate (30S RNA) was the same as that observed in 18S and 28S mature rRNA (67-75%), indicating that synthesis of 40S pre-rRNA is inhibited, but the processing of 40S pre-rRNA is not.

We know that 30-60 min is needed for the pulse-labeled 40 S pre-rRNA to be processed into 18S and 28 S rRNAs. We, therefore, first pulse-labeled neurula cells for 35 min and then treated the cells with ammonium chloride or trimethylammonium perchloride in the fresh, ^{3}H-uridine-free, medium that contained actinomycin D (10 μg/ml) as a transcription inhibitor. When neurula cells were pulse-labeled for 35 min, they contained only 40 S pre-rRNA and heterogeneous mRNA-like RNA, and when they were further cultured for 2 hrs in the fresh medium that did not contain ^{3}H-uridine but contained actinomycin D, the label in the 40 S pre-rRNA was quantitatively recovered in 18 S and 28S mature rRNAs. We found that such processing occurred also in the presence of 10 mM ammonium chloride or trimethylammonium perchloride. Therefore, we concluded that weak bases inhibited rRNA transcription but did not inhibit the processing of 40 S pre-rRNA into two mature rRNAs. The almost complete recovery of the label in 40S pre-rRNA in the two mature rRNAs again supports the absence of aberrant processing of rRNA in the presence of weak bases.

9. Cell Physiological Studies of the Mechanism of the Inhibition of rRNA Gene Expression by Weak Bases

As for the mode of action of weak bases, we suspected that a slight elevation of intracellular pH might be involved, since such pH change has previously been demonstrated as a mode of the action of weak bases (Webb and Charbonneau, 1986). In such a situation, it is known that removal of Na^{+} from the medium abolish the slight elevation of intracellular pH, thereby abolishing the pH-mediated effect of weak bases (Stith and Maller, 1985;

Wasserman et al., 1984). We, therefore, tested the effect of 3 mM ammonium chloride and 1.5 mM trimethylammonium perchloride using *Xenopus* neurula cells in a medium in which all the Na^+ was replaced by choline ions. Results of the labeling experiment showed that the inhibitory effects on rRNA synthesis of both ammonium chloride and trimethylammonium perchloride were completely abolished. These results strongly suggest that for the inhibitory effect of both of these weak bases to be realized Na^+ is needed in the culture medium. The next question here is whether or not we could actually detect the pH increase in the cells treated with weak bases, and this issue is now being explored in our laboratory.

We also measured the level of ATP and other ribonucleotides in the acid-soluble fraction of neurula cells after treating them for 3 hrs with 3 mM ammonium chloride or 1.5 mM trimethylammonium perchloride. The results obtained showed that the level of ATP as well as other ribonucleotides triphosphates remained unchanged. The amount of the ATP detected in one-embryo equivalent neurula cells in the control culture was about 1000 pmoles, which is consistent with our previous determination (Tashiro et al., 1983). Therefore, neither ammonium chloride nor trimethylammonium perchloride appears to disturb the energy generating system in *Xenopus* neurula cells.

10. Ammonia as a Candidate for One of the Factors Involved in the Control of rRNA Gene Expression during *Xenopus* Embryogenesis

When we examined ninhydrin-positive materials in the acid-soluble fraction of *Xenopus* early embryos, there was a sizable amount of ammonia but not amines within the embryo. When we performed similar analyses at various stages of development, we found that the level of ammonia in an egg was ca. 55 ng/egg before fertilization, which roughly corresponds to ca. 3 mM of ammonia as an intra-egg concentration (the volume of an egg is ca. 1µl). During cleavage, this level did not change greatly, but after MBT it decreased to the level as low as ca. 15 ng/embryo (or ca. 1 mM as an intra-embryo concentration) and the lowered level was maintained throughout later stages (Figure 9).

Since we suspected that ammonia might actually be involved in the regulation of rRNA gene expression in *Xenopus* embryos, we attempted to isolate ammonia (and amines if any) from cleavage stage embryos. Starting from 25,000 cleavage stage embyos, we isolated about 10 mg of residual materials after evaporating the hydrochloric acid solution that was used to capture the cellular volatile ammonia components. Mass spectrometric analysis revealed the presence of ammonium chloride but not of amines, reconfirming that early cleavage stage embryos contain ammonia but not amines. Ten milligrams of ammonium chloride obtained here corresponded to ca. 4 mg of ammonia, which is comparable to the amount of ammonia (ca. 2 mg) that was expected to be present in the 25,000 cleavage stage embryos on the basis of the data in Figures 8. We tested effects of this cleavage stage embryo-derived ammonium chloride on rRNA synthesis in neurula cells, with a result that rRNA synthesis was inhibited by 60% and 90% with only a slight inhibition (less than 10 %) in 4S RNA (tRNA) synthesis in the presence of 1.0 and 5.0 mM of this ammonium chloride preparation.

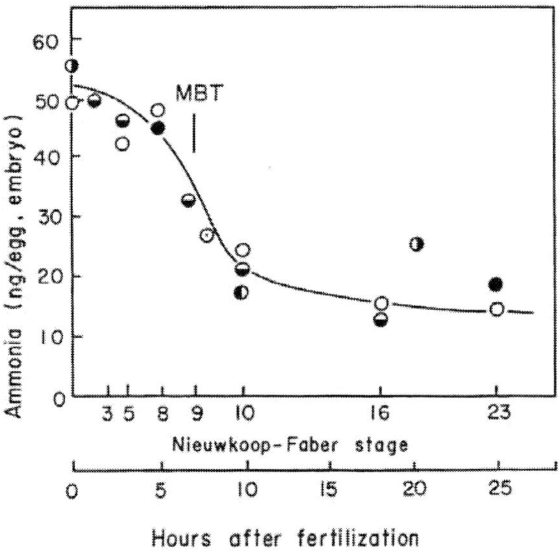

Figure 9. Changes in the amount of ammonium ion in eggs and embryos of *Xenopus laevis* during development. Two hundreds of embryos at different stages were homogenized with 2 ml of 10% TCA (trichloroacetic acid) and amount of ammonium ion determined in an amino acid analyzer as in Fig. 7. Embryos were cultured in 0.1 X Steinberg's solution. Different symbols are for different batches of embryos. From Shiokawa et al (1986).

CONCLUSION AND A WORKING HYPOTHESIS

Using *Xenopus* embryonic cells, we studied rRNA synthesis, and eventually found that weak bases, especially ammonium ion selectively inhibits rRNA synthesis. We described here about several characteristic aspects of *Xenopus* embryos as an experimental materials and to this was added a brief description about maternally preset program of apoptosis found as a fail-safe mechanism of early development. Then, in our efforts to isolate substances that control the initiation and activation of rRNA synthesis in *Xenopus* embryos, we happened to obtain data that conditioned medium of blastula cells was slightly inhibitory on rRNA synthesis in isolated neurula cells. Since boiled conditioned medium had similar effects, we expected that the active substance might be a low-molecular-substance. However, the results with the conditioned medium were not highly reproducible, and therefore, we proceeded to directly prepare a perchloric acid-soluble fraction of early embryos, and tried to isolate the putative inhibitory substance from the homogenate. One possible reason for the low reproducibility of the results at the initial step of the experiment may be that in the initial step of our experiment 4S RNA synthesis was detected by counting the total radioactivity which was eluted from the MAK column by 0.4 M NaCl in a stepwise manner. But, anyway, we moved to analyze PCA-extract of early embryo homogenates.

PCA-soluble materials were neutralized with KOH and adsorbed on a charcoal column and eluted using ammonia-alcohol (0.2 N ammonia-50% ethanol). The ammonia-alcohol eluate was constantly active as the inhibitor. But, it turned out that the active substance in the charcoal eluate is ammonium perchlorate which must have been formed during chromatography. We then started to examine the effects of various amines and ammonium salts on rRNA synthesis, and finally reached a conclusion that ammonium ion could be a regulative factor of rRNA synthesis. The inhibition of rRNA gene expression is due to the inhibition of 40 S pre-rRNA transcription. Furthermore, the inhibition of rRNA synthesis appeared to be induced by a slight increase in the intracellular pH, since in the Na^+-lacking medium ammonium ion became unable to inhibit rRNA synthesis. The increase in the level of intracellular ammonia may induce transamination of oxaloacetic acid into asparric acid and this could result in inhibition of TCA cycle. However, this did not seem to be the case, a finding consistent with the fact that various amines which do not bind to oxaloaceticacid are also effective in inhibiting rRNA synthesis.

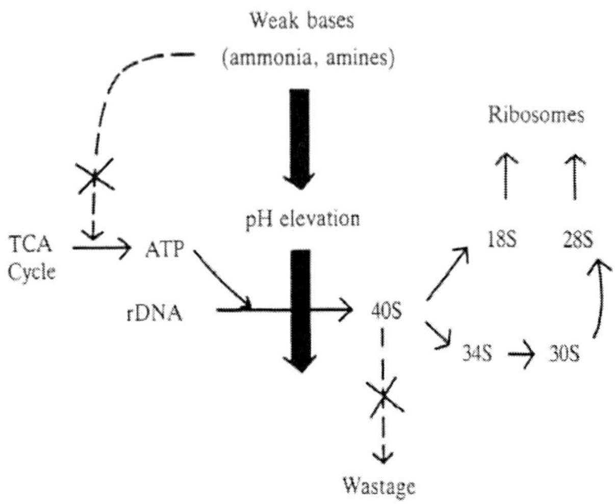

Figure 10. A working hypothesis on the mode of action of weak bases on rRNA gene expression. Weak bases do not interfere with ATP generation (upper X) and do not induce degradation of 40S pre-rRNA (lower X), but inhibit 40S pre-rRNA formation (lower large arrow), probably at the transcription level *via* a slight elevation of intrcellular pH (upper large arrow). From Shiokawa et al. (1987).

In *Xenopus*, a slight increase of pH is most likely due to changes in the levels of weak bases. This has been implicated as a necessary step in oocyte maturation (Stith and Maller, 1984) and in fertlization (Nuccitelli et al., 1981). Furthermore, it has been reported that cleavage of fertilized *Xenopus* eggs commences when the previously elevated pH is lowered with weak acids (Lee and Steinhardt, 1981). Thus, we propose here a working hypothesis that amnmonia suppresses rDNA transcription via a process of a slight increase in the intracellular pH (Figure 10).

References

Andeol YA (1994) Early transcription in different animal species: Implication for transition from maternal to zygotic control in development. *Roux's Arch. Dev. Biol.*, 204: 3-10.

Brown DD, Caston JD (1962) Biochemistry of amphibian development. I. Ribosome and protein synthesis in early developmet of *Rana pipiens*. *Develop. Biol*, 5: 412-434.

Brown DD, Dawid IB (1968) Specific gene amplification in oocytes. *Science,* 160: 272-280.

Brown DD, Littna E (1964) RNA synthesis during the development of *Xenopus* laevis, the African clawed toad. *J. Mol. Biol.*, 8: 669-687.

Brown DD, Littna E (1966) Synthesis and accumulation of DNA-like RNA during embryogenesis of *Xenopus laevis*. *J. Mol. Biol.*, 20: 81-94.

Busby SJ, Reeder RH (1983) Spacer sequences regulate transcription of ribosomal gene plsmids injected into *Xenopus* embryos. *Cell*, 34: 989-996.

Carter AD, Sible JC (2003) Loss of XChk1 function triggers apoptosis after the midblastula transition in *Xenopus* embryos. *Mech. Dev.*, 120: 315-323.

Cohen S (1954) The mtabolism of $C^{14}O_2$ during amphibian development. *J. Biol. Chem.*, 211: 337-354.

Emanuelsson H, Heby O (1978) Inhibition of putrescine synthesis blocks development of the polychete *Ophryotrocha labronica* at gastrulation. *Proc. Natl. Acad. Sci. USA*, 75: 1039-1042.

Gurdon JB (1974) The control of gene expression in animal development. *Clarendon Press*, Oxford pp79.

Gurdon JB, Brown DD (1965) Cytoplasmic regulation of RNA synthesis and nucleolar formation in developing embryos *Xenopus laevis*. *J. Mol. Biol.*, 12: 27-35.

Heasman J (2006) Patterning the early *Xenopus* embryo. *Development,* 133: 1205-1217.

Kai M, Higo T, Yokoska J,Kaito C, Fukamachi H, Takayama E, Igarashi K, Shiokawa K (2000) Overexpression of *S*-adenosylmethionine decarboxylase (SAMDC) activates the maternal program of apoptosis shortly after MBT in *Xenopus* embryos. *Int. J. Dev. Biol.*, 44: 507-510.

Kai M, Kaito C, Fukamachi H, Higo T, Takayama E, Hara H, Ohya Y, Igarashi K, Shiokawa K (2003) Overexpression of S-adenosylmethionine decarboxylase (SAMDC) in *Xenopus* embryos activates maternal program of apoptosis as a "fail-safe" mechanism of early embryogenesis. *Cell Res.*, 13: 147-158.

Kaito C, Kai M, Higo T, Takayama E, Fukamachi H, Sekimizu K, Shiokawa K (2001) Activation of the maternally preset program of apoptosis by microinjection of 5-aza-2'-deoxycytidine and 5-methyl-2'-deoxycytidine-5'-triphosphate in *Xenopus laevis* embryos. *Develop. Growth Differ.*, 43: 383-390.

Kajita E, Wakiyama M, Miura K-I, Mizumoto K, Oka T, Komuro I, Miyata T, Yatsuki H, Hori K, Shiokawa K (2000). Isolation and characterization of Xenopus laevis aldolase B cDNA and expression pattern of aldolase A, B, and C genes in adult tissues, oocytes and embryos of *Xenopus laevis*. *Biochim. Biophys. Acta*, 1493: 101-118.

Laskey RA, Gerhart J, Knowland JS (1973) Inhibition of ribosomal RNA synthesis in neurula cells by extracts from blastulae of *Xenopus laevis*. *Devel. Biol.*, 33: 241-247.

Lee, SC, Steinhardt, RA (1981) Observation on intracellular pH during cleavage of eggs of *Xenopus laevis*. *J. Cell Biol.*, 91: 414-419.

Li P, Nijhawan D, Budihardjo I, Srinivasula SM, Ahmad M, Alnemri ES, Wang S (1997) Cytochrome C and dATP-dependent formation of Apaf-1/caspase-9 complex initiates an apoptotic cascade. *Cell*, 91: 479-489.

Lieberman I, Abrams R, Ove P (1963) Changes in the metabolism of ribonucleic acid preceding the synthesis of deoxyribonucleic acid in mammalian cells cultured from the animal. *J. Biol. Chem.*, 238: 2141-2149.

Minoura I, Nakamura H, Tashiro K, Shiokawa K (1995) Stimulation of circus movement by activin, bFGF and TGF-beta 2 in isolated animal cap cells of *Xenopus laevis*. *Mech. Dev.*, 49: 65-69.

Nakahashi T, Yamana K (1976) Biochemical and cytological examination of the examination of the initiation of ribosomal RNA synthesis during gastrulation of *Xenopus laevis*. *Develop. Groth Differ.*, 18: 329-339.

Nakakura N, Miura T, Yamana K, Ito A, Shiokawa K (1987) Synthesis of heterogeneous mRNA-like RNA and low-molecular-weight RNA before the midblastula transition in embryos of *Xenopus laevis*. *Dev. Biol.*, 123: 421-429.

Newport J, Kirschner M (1982) A major developmental transition in early *Xenopus* embryos: I. Characterization and timing of cellular changes at the midblastula stage. *Cell,* 30: 675-86.

Nieuwkoop PD, Faber J (1974) Normal Table of *Xenopus laevis* (Daudin). North-Holland Publ.Co., Amsterdam.

Nuccitelli R, Webb DJ, Lagier ST, and Matson GB (1981) ^{31}P NMR reveals increased intracellular pH after fertilization in *Xenopus* eggs. *Proc. Natl. Acad. Sci. USA*, 78: 4421-4425.

Richter JD, Wasserman WJ, Smith LD (1982) The mechanism for increased protein synthesis during *Xenopus* oocyte maturation. *Dev. Biol.*, 89: 159-167.

Russell DH (1983) Microinjection of purified ornithine decarboxylase into *Xenopus* oocytes selectively stimulates ribosomal RNA synthesis. *Proc. Natl. Acad. Sci. USA*, 80: 1318-1321.

Shibata M, Shinga J, Yasuhiko Y, Kai M, Miura K-I, Shimogori T, Kashiwagi K, Igarashi K, Shiokawa K (1998) Overexpression of S-adenosylmethionine decarboxylase (SAMDC) in early *Xenopus* embryos induces cell dissociation and inhibits transition from the blastula to gastrula stage. *Int. J. Dev. Biol.*, 42: 675-686.

Shinga J, Kashiwagi K, Tashiro K, Iigarashi K, Shiokawa K (1996) Maternal and zygotic expression of mRNA for *S*-adenosylmethionine decarboxylase and its relevance to the unique polyamine composition in *Xenopus* oocytes and embryos. *Biochim. Biophys. Acta*, 1308: 31-40.

Shiokawa K, Aso M, Kondo R, Takai J, Yoshida J, Mishina T, Fuchimukai K, Ogasawara T, Kariya T, Tashiro K, Igarashi K (2010). Effects of S-adenosylmethionine decarboxulase, polyamines, amino acids, and weak bases (amines and ammonia) on development and ribosomal RNA synthesis in *Xenopus* embryos. *Amino Acids*, 38: 439-449.

Shiokawa, K., Aso, M., Kondo, T., Uchiyama, H., Kuroyanagi, S., Takai, J., Takahashi, S., Kajitani, M., Kaito, C., Sekimizu, K., Takayama, E., Igarashi, K., and Hara, H. 2008. Gene expression in pre-MBT embryos and activation of maternally-inherited program of apoptosis to be executed at around MBT as a fail-safe mechanism in *Xenopus* early embryogenesis. *Gene Regulation Systems Biol.*, 2: 1-18.

Shiokawa K, Kai M, Higo T, Kaito C, Yokoska J, Yasuhiko Y, Kajita E, Nagano M, Yamada Y, Shibata M, Nuto T, Shinga J, Hara H, Takayama E, Fukamachi H, Yaoita Y, Igarashi K. (2000). Maternal propgram of apoptosis activated shortly after midblastula transition by overexpression of S-adenosylmethionine decaroboxylase in *Xenopus* early embryos. *Compar. Biochem. Physio. Part B*, 126: 149-155.

Shiokawa K, Kawazoe Y, Nomura H, Miura T, Nakakura N, Horiuchi T, and Yamana K (1986a) Ammonium ion as a possible regulator of the commencement of rrna synthesis in *Xenopus laevis* embryogenesis. *Dev. Biol.*, 115: 380-391.

Shiokawa, K., Kawazoe, Y., Tashiro, K., and Yamana, K (1986b) Effects of various ammonium salts, amines, polyamines, and alpha-methylornithine on rRNA synthesis in neurula cells of *Xenopus laevis* and *Xenopus borealis*. *Cell Differ.*, 18: 101-108.

Shiokawa K, Kawazoe Y, and Yamana K (1985) Demonstration that inhibitor of rRNA synthesis in "charcoal-extracts" of Xenopus embryos is artifactually produced ammonium perchlorate. *Dev. Biol.*, 112: 258-260.

Shiokawa K, Kurashima R, Shinga J (1994) Temporal control of gene expression from endogenous and exogenously-introduced DNAs in early embryogenesis of *Xenopus laevis*. *Int. J. Dev. Biol.*, 38: 249-255.

Shiokawa K, Misumi Y, Yamana K (1981a) Demonstration of rRNA synthesis in pre-gastrular embryos of *Xenopus laevis*. *Develop. Growth Differ.*, 23: 579-587.

Shiokawa K, Misumi Y, Tashiro K, Nakakura N, Yamana K, Oh-uchida M (1989) Changes in the patterns of RNA synthesis in early embryogenesis of *Xenopus laevis*. *Cell Differ. Dev.*, 28: 17-25.

Shiokawa K, Misumi Y, Yasuda Y, Nishio Y, Kurata S, Sameshima M, Yamana K (1979) Synthesis and transport of various RNA species in developing embryos of Xenopus laevis. *Dev. Biol.*, 68; 503-514.

Shiokawa k, Nada O, Yamana K (1967) RNA synthesis in isolated cells from Xenous laevis embryos. *Nature*, 213: 1027-1028.

Shiokawa K, Takayama E, Higo T, Kuroyanagi S, Kaito C, Hara H, Kajitani M, Sekimizu K, Tadakuma T, Miura K-I, Igarashi K, Yaoita Y (2005) Occurrence of pre-MBT synthesis of caspase-8 mRNA and activation of caspase-8 prior to execution of SAMDC (S-adenosylmethionine decarboxylase)-induced, but not p53-induced, apoptosis in *Xenopus* late blastulae. *Biochem. Biophys. Res. Commun.*, 336: 682-691.

Shiokawa K, Tashiro K, Misumi Y, Yamana K (1981b) Non-coordinated synthesis of RNAs in pre-gastrular embryos of Xenopus laevis. *Develop. Growth Differ.*, 23: 589-597.

Shiokawa K, Yamana K (1965) Demonstration of "polyphosphate" and its possible role in RNA synthesis during early development of *Rana japonica* embryos. *Exptl. Cell Res.*, 38: 180-186.

Shiokawa K, Yamana K (1967a) Pattern of RNA synthesis in isolated cells of *Xenopus laevis* embryos. *Dev. Biol.*, 16: 368-388.

Shiokawa K, Yamana K (1967b) Inhibitor of ribosomal RNA synthesis in Xenopus laevis embryos. *Devel. Biol.*, 16: 389-406.

Stith B J, Maller JL (1984) The effect of insulin on intracellular pH and ribosomal protein S6 phosphorylation in oocytes of Xenopus laevis. *Dev. Biol.*, 102: 79-89.

Stith BJ, Maller JL (1985) Increased intracellular pH is not necessary for ribosomal protein S6 phosphorylation increased protein synthesis or germinal vesicle breakdown in Xenopus oocytes. *Develop. Biol.*, 107: 460-469.

Takayama E, Higo T, Kai M, Fukasawa M, Nakajima K, Hara H, Tadakuma T, Igarashi K, Yaoita Y, Shiokawa K (2004) Involvement of caspase-9 in execution of the maternal program of apoptosis in *Xenopus* late blastulae overexpressed with S-adenosylmethionine decdarboxylase. *Biochem. Biophys. Res. Commun.*, 325: 1367-1375.

Tashiro K, Misumi Y, Shiokawa K, Yamana K (1983) Determination of the rate of rrna synthesis in xenopus laevis triploid embryos produced by low-temperature treatment. *J. Exp. Zool.*, 225: 489-495.

Wasserman WJ, Houle JG, Samuel D (1984) The maturation response of stage IV, V, VI *Xenopus* oocytes to progesterone stimulation *in vitro*. *Dev. Biol.*, 105: 315-324.

Webb DJ, Charbonneau M (1986) Weak bases inhibit cleavage and embryogenesis in amphibians, Xenopus (toad) and Pleurodels (newt) and echinoderms Paracentrotus sphaerechinus (sea urchin) and asterias (starfish). *Cell Differ.*, 20: 33-44.

Wroble BN, Sible JC (2005) Chk2/Cds1 protein kinase blocks apoptosis during early development of *Xenopus laevis*. *Dev. Dyn.*, 233: 1359-1365.

Yang J, Tan C, Darken RS, Wilson PA, Klein PS (2002) β-Catenin/Tcf regulated transcription prior to the midblastula transition. *Development,* 129: 5743-5752.

Yasuda GK, Schubiger G (1992) Temporal regulation in the early embryo: Is MBT too good to be true? *Trends Genet.*, 8: 124-127.

In: Ammonia: Structure, Biosynthesis... ISBN: 978-1-62100-502-5
Editors: V.A. Fekete, et al, pp. 91-98

Chapter 4

PLANT ABIOTIC STRESS RESPONSES AND NUTRIENTS

Yuriko Osakabe[1] **and Keishi Osakabe**[2]*

[1] Graduate School of Agricultural and Life Sciences, University of Tokyo, 1-1-1 Yayoi, Bunkyo-ku, Tokyo 113-8657.

[2] Department of Plant Biotechnology, National Institute of Agrobiological Sciences, 2-1-2 Kannondai, Tsukuba, Ibaraki 305-8602.

ABSTRACT

Plants absorb nitrogen as nitrate or ammonium ions from the soil, and the nitrogen is assimilated into the amino acids. Under the environmental stress conditions, plant assimilation of nitrogen and the metabolic pathway in amino acid biosynthesis can be affected. In drought and salinity conditions, amino acids, such as proline, accumulate and function as osmolytes that affect osmotic adjustment in plant cells. Proline synthesis also affects the biogenesis of reactive oxygen species (ROS). Phenylalanine is synthesized with glutamate and converted to *trans*-cinnamic acid by phenylalanine ammonia-lyase (PAL), which catalyzes the first reaction in phenylpropanoid metabolism. The expression of a variety of genes that function in the metabolic pathway to increase stress tolerance is upregulated in plant cells. In this chapter, we present nitrogen assimilation under stress conditions and focus on the

* Fax: +81-3-5841-8136, e-mail: ayosa@mail.ecc.u-tokyo.ac.jp

transcriptome and metabolome studies in regulatory networks in plant abiotic stress tolerance.

Keywords: plant; nitrogen; abiotic stress; proline; phenylalanine; reactive oxygen species.

INTRODUCTION

Ammonia is a source of nitrogen for plants and is assimilated into amino acids and proteins. Nitrogen is an essential nutrient as well as an important factor for agricultural productivity. Nitrogen availability regulates various plant growth processes and environmental conditions. The adaptations in the metabolic pathway in response to nitrogen supply, including the changes in the expression patterns of various functional genes, are observed in plants. It has been proved that amino acids, such as proline, accumulate and function as osmolytes in plant cells during osmotic stress (Szabados and Savouré 2010). Phenylalanine is converted to *trans*-cinnamic acid by phenylalanine ammonia-lyase (PAL), which is a key enzyme in phenylpropanoid metabolism. Phenylpropanoid compounds are accumulated by various environmental stresses, such as UV radiation, low temperatures, pathogen attack, wounding, and oxidative and drought stresses, and they function as antioxidants (Dixon and. Paiva 1995; Osakabe et al. 2009a; 2009b; 2011).

Recently, the metabolic changes in response to environmental stresses, such as drought, salinity, and cold stresses, have been identified by integrated metabolome and transcriptome analysis (Urano et al., 2010). These findings suggested that the coordinated regulatory networks control plant physiological responses to the stress. In this chapter, we summarize the amino acids that play roles in plant adaptation to abiotic stress.

REGULATION OF PROLINE ACCUMULATION IN PLANTS

Proline has been considered an osmolyte, and it functions as a molecular chaperone that protects subcellular structures, proteins, peptides, and nucleic acids under osmotic stress. Recent studies have shown that proline may also play the role of an antioxidant in stress tolerance to affect redox homeostasis. Application of proline causes a reduction in ROS levels of fungi and yeast and represses programmed cell death (Chen and Dickman 2005). Proline also plays an important role in ROS scavenging in human cells, in which it suppresses ROS and protects the cells against carcinogenic oxidative stress (Krishnana et al., 2008). The overexpression of Δ^1-pyrroline-5-carboxylate synthetase (P5CS), which is a key enzyme of proline biosynthesis (Figure 1), in transgenic tobacco plants and algae resulted in a high accumulation of proline as well as a reduction in the levels of free radicals in the cells (Hong et al., 2000; Sirípornadulsil et al., 2002). The T-DNA insertion mutants, *p5cs*, resulted in a reduction of stress-induced proline, ROS accumulation, and hypersensitivity to salt stress (Székely, et al., 2008). The *p5cs* exhibited hypersensitivity to salt stress, due to the reduced activity of key antioxdant enzymes, thereby leading to a high accumulation of ROS (Székely, et al., 2008). Thus, the relationship between photosynthesis and proline metabilism under stress has been proved. The inhibition of proline biosynthesis by the antisense *P5CR* (*Δ^1-pyrroline-5-carboxylate reductase*) gene (Figure 1) in the transgenic soybean plants caused a reduction of NADPH to $NADP^+$ reaction and resulted in drought hypersensitivity (De Ronde et al., 2004). Proline biosynthesis requires NADPH for the reduction of glutamate to Δ^1-pyrroline-5-carboxylate (P5C) and of P5C to proline, and it generates $NADP^+$ (Figure 1). Proline biosynthesis that occurs in plastids during stress maintains low levels of NADPH: $NADP^+$ and affects the redox balance.

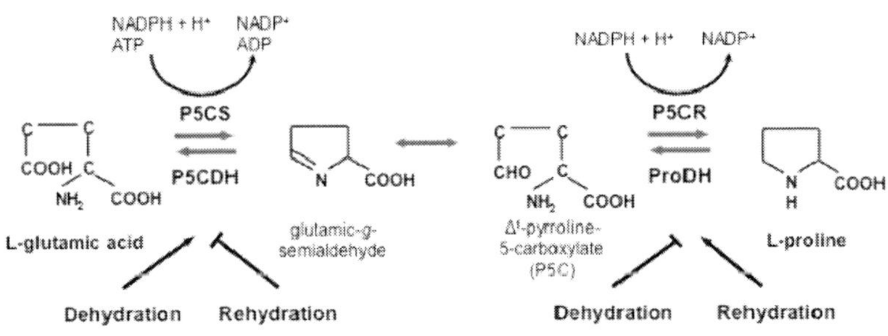

Figure 1. Biosynthesis and metabolism of proline in plants. Proline biosynthesis is activated by drought stress, whereas proline catabolism is repressed by drought stress. P5CS, Δ^1-pyrroline-5-carboxylate synthetase; P5CR, Δ^1-pyrroline-5-carboxylate reductase; P5CDH, Δ^1-Pyrroline-5-carboxylate dehydrogenase; ProDH, proline dehydrogenase.

Proline biosynthesis is activated by drought stress, whereas proline catabolism is repressed by the stress (Figure 1, 2). *Arabidopsis P5CS1* gene is induced by osmotic and salt stresses and abscisic acid (ABA) (Yoshida et al. 1995; Strizhov et al. 1997; Savoure et al. 1997; Verslues et al., 2007). On the contrary, *ProDH* (proline dehydrogenase) transcription levels are activated by rehydration and proline accumulation, but repressed by dehydration stress (Kiyosue et al. 1996; Verbruggen et al. 1996). These studies suggest that transcriptional control is one of the important factors in feedback inhibition of proline metabolism. A *cis*-acting element, ACTCAT, for the *ProDH* expression in the proline and hypo-osmolarity-responsivity has been found (Figure 2), and the basic leucine zipper protein (bZIP) transcription factors (AtbZIP2, AtbZIP11, AtbZIP44, and AtbZIP53) have been identified as regulator proteins in its activation (Satoh et al. 2002; 2004).

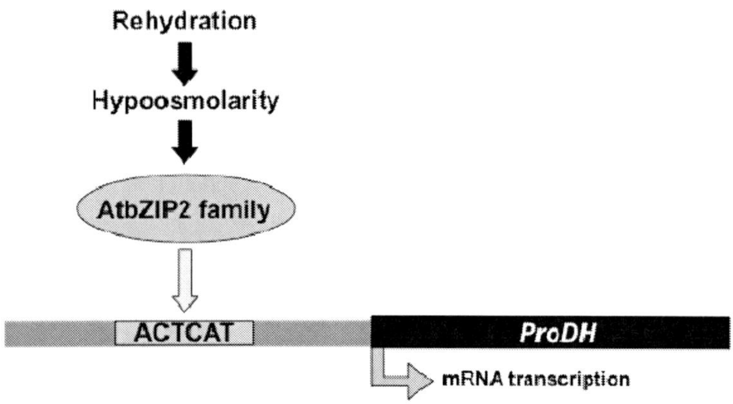

Figure 2. A model for the induction of *ProDH* under hypo-osmolarity. *ProDH* transcription levels are activated by rehydration and proline accumulation, but repressed by dehydration stress. A *cis*-acting element, ACTCAT, in the *ProDH* promoter controls its expression under the hypo-osmolarity condition, and the transcription factors, AtbZIP2, AtbZIP11, AtbZIP44, and AtbZIP53, have been identified as regulator proteins in the activation of ACTCAT.

Phenylpropanoids in Plant Stress Responses

Phenylalanine is converted to *trans*-cinnamic acid by phenylalanine ammonia-lyase (PAL), which is a key enzyme in phenylpropanoid metabolism. Phenylpropanoid compounds, which are derived from phenylalanine, are accumulated by various environmental stresses, such as UV radiation, low temperatures, pathogen attack, wounding, and oxidative and drought stresses, and they function as antioxidants (Dixon and Paiva 1995; Osakabe et al. 2009a; 2009b; 2011). Many studies indicate that various enzymes in phenylpropanoid biosynthesis are controlled in response to various environmental conditions and developmental stages (Boudet et al. 2003, Boerjan et al. 2003, Osakabe et al. 1999, Raes et al. 2003). The phenylpropanoid biosynthesis is controlled by a spatial and temporal regulatory mechanism. Lignin is a complex phenolic polymer as well as a major component in the secondary cell wall. Lignin provides hydrophobicity and rigidity to the thickening cell wall for water transport and mechanical support. Lignifications and biosynthesis of lignin precursors are induced by biotic and abiotic stresses, such as wounding, pathogen attack, and drought stresses (Vance et al. 1980, Dixon and Paiva 1995, Alvarez et al. 2008, Hu et al. 1999). Phenylpropanoids are also the precursors for a wide range of secondary metabolites, such as anthocyanin, flavonoids, isoflavonoids, cumarins, and stilbens. These compounds play important roles in defense responses against pathogens, protectants from UV light, and in signal transduction with other organisms (Ferrer et al. 2008; Huang et al., 2010).

A recent study on four PAL genes, *PAL1*, *PAL2*, *PAL3*, and *PAL4*, in *Arabidopsis* indicated distinct as well as overlapping roles of the *Arabidopsis PAL* genes in growth, development, and responses to environmental stresses (Huang et al., 2010). In this study, the *pal1 pal2* double mutants generated yellow seeds resulting from the lack of condensed tannin in the seed coat and were also deficient in anthocyanin accumulation in various tissues that accumulate in wild-type plants during stress, thus suggesting that these *PAL* genes have redundant roles in flavonoid biosynthesis. The *pal1 pal2* mutants were more sensitive to UV-B light. The *pal1 pal2 pal3 pal4* quadruple knockout mutants showed reduced levels of lignin content and plant growth. They also accumulated reduced levels of the plant hormone salicylic acid (SA) and displayed increased susceptibility to the virulent strain *Pseudomonas syringae*, thus suggesting that PAL is a key enzyme in SA biosynthesis in *Arabidopsis* (Huang et al., 2010).

Pheylpropanoids play roles as antioxidants and possess free-radical-scavenging properties (Korkina 2007). The phenolic compounds from plants are also biologically active components and are used as valuable chemicals for human health. For medicinal purpose, plant phenylpropanoids, such as UV screens and anti-pathogen and anti-tumor chemicals, are used as antioxidants. In conclusion, recent studies are focused on metabolic engineering and identifying novel characteristics of nitrogen assimilation including amio acid biosynthesis and phenylpropanoid metabolism during abiotic stress. These studies will provide potential benefits to human health as well as to plant productivity.

References

Alvarez S, Marsh EL, Schroeder SG, Schachtman DP (2008) Metabolomic and proteomic changes in the xylem sap of maize under drought. *Plant Cell Environ.* 31: 325-340.

Boerjan W, Ralph J, Baucher M (2003) Lignin biosynthesis. *Ann. Rev. Plant Biol.* 54: 519-546.

Boudet AM, Kajita S, Grima-Pettenati J, Goffner D (2003) Lignins and lignocellulosics: a better control of synthesis for new and improved uses. *Trends Plant Sci.* 8: 576-581.

Chen C, Dickman MB (2005) Proline suppresses apoptosis in the fungal pathogen Colletotrichum trifolii. *Proc. Natl. Acad. Sci. USA* 102: 3459-3464.

De Ronde JA, Cress WA, Krüger GH, Strasser RJ, Van Staden J. (2004) Photosynthetic response of transgenic soybean plants, containing an Arabidopsis P5CR gene, during heat and drought stress. *J. Plant Physiol.* 161: 1211-1224.

Dixon RA, Paiva NL (1995) Stress-lnduced Phenylpropanoid Metabolism. *Plant Cell* 7: 1085-1097.

Ferrer JL, Austin MB, Stewart Jr C, Noel JP (2008) Structure and function of enzymes involved in the biosynthesis of phenylpropanoids. *Plant Physiol. Biochem.* 46: 356-370.

Hong Z, Lakkineni K, Zhang Z Verma DPS (2000) Removal of feedback inhibition of 1-pyrroline-5-carboxylate synthetase results in increased proline accumulation and protection of plants from osmotic stress. *Plant Physiol.* 122:1129-1136.

Hu WJ, Harding SA, Lung J, Popko JL, Ralph J, Stokke DD, Tsai CJ, Chiang VL. (1999) Repression of lignin biosynthesis promotes cellulose accumulation and growth in transgenic trees. *Nat. Biotech.* 17: 808-812.

Huang J, Gu M, Lai Z, Fan B, Shi K, Zhou YH, Yu JQ, Chen Z (2010) Functional analysis of the Arabidopsis PAL gene family in plant growth, development, and response to environmental stress. *Plant Physiol.* 153: 1526-1538.

Kiyosue T, Yoshiba Y, Yamaguchi-Shinozaki K, Shinozaki K (1996) A nuclear gene encoding mitochondrial proline dehydrogenase, an enzyme involved in proline metabolism, is upregulated by proline but downregulated by dehydration in Arabidopsis. *Plant Cell* 8: 1323-1335.

Korkina LG (2007) Phenylpropanoids as naturally occurring antioxidants: From plant defense to human health. *Cellular and Molecular Biology* 53: 15-25.

Krishnana N, Dickmanb MB, Becker DF (2008) Proline modulates the intracellular redox environment and protects mammalian cells against oxidative stress. *Free Radical Biology and Medicine* 44, 671-681.

Osakabe K, Tsao CC, Li L, Popko JL, Umezawa T, Carraway DT, Smeltzer RH, Joshi CP, Chiang VL. (1999) Coniferyl aldehyde 5-hydroxylation and methylation direct syringyl lignin biosynthesis in angiosperms. *Proc. Natl. Acad. Sci. USA* 96: 8955-8960.

Osakabe Y, Osakabe K, Chiang VL (2009a) Isolation of 4-coumarate Co-A ligase gene promoter from loblolly pine (*Pinus taeda*) and characterization of tissue-specific activity in transgenic tobacco. *Plant Physiol. Biochem.* 47: 1031-1036.

Osakabe Y, Osakabe K, Chiang VL (2009b) Characterization of the tissue-specific expression of phenylalanine ammonia-lyase gene promoter from loblolly pine (*Pinus taeda*) in *Nicotiana tabacum*. *Plant Cell Rep.* 28: 1309-1317.

Osakabe Y, Kajita S, Osakabe K (2011) Genetic engineering of woody plants: current and future targets in a stressful environment. *Physiol. Plant* 142:105-117.

Raes J, Rohde A, Christensen JH, Van de Peer Y, Boerjan W (2003) Genome-wide characterization of the lignification toolbox in Arabidopsis. *Plant Physiology* 133: 1051-1071.

Satoh R, Nakashima K, Seki M, Shinozaki K, Yamaguchi-Shinozaki K (2002) ACTCAT, a novel *cis*-acting element for proline- and hypoosmolarity-responsive expression of the *ProDH* gene encoding proline dehydrogenase in Arabidopsis. *Plant Physiol.* 130: 709-719.

Satoh R, Fujita Y, Nakashima K, Shinozaki K, Yamaguchi-Shinozaki K (2004) A novel subgroup of bZIP proteins functions as transcriptional activators in hypoosmolarity-responsive expression of the *ProDH* gene in Arabidopsis. *Plant Cell Physiol.* 45: 309-317.

Savouré A, Hua XJ, Bertauche N, Van Montagu M, Verbruggen N (1997) Abscisic acid-independent and abscisic acid-dependent regulation of the proline biosynthesis upon cold and osmotic stresses in Arabidopsis thaliana. *Mol. Gen. Genet.* 254:104-109.

Siripornadulsil S, Traina S, Verma DP, Sayre RT (2002) Molecular mechanisms of proline-mediated tolerance to toxic heavy metals in transgenic microalgae. *Plant Cell* 14 :2837-2847.

Strizhov N, Abraham E, Okresz L, Blickling S, Zilberstein A, Schell J, Koncz C, Szabados L (1997) Differential expression of two P5CS genes controlling proline accumulation during salt-stress is regulated by ABA1, ABI1, and AXR2 in *Arabidopsis*. *Plant Journal* 12: 557–569.

Szabados L, Savouré A (2010) Proline: a multifunctional amino acid. *Trends Plant Sci.* 15: 89-97.

Székely G, Abrahám E, Cséplo A, Rigó G, Zsigmond L, Csiszár J, Ayaydin F, Strizhov N, Jásik J, Schmelzer E, Koncz C, Szabados L (2008) Duplicated *P5CS* genes of Arabidopsis play distinct roles in stress regulation and developmental control of proline biosynthesis. *Plant J.* 53: 11-28.

Urano K, Maruyama K, Ogata Y, Morishita Y, Takeda M, Sakurai N, Suzuki H, Saito K, Shibata D, Kobayashi M, Yamaguchi-Shinozaki K, Shinozaki K (2009) Characterization of the ABA-regulated global responses to dehydration in Arabidopsis by metabolomics. *Plant J.* 57: 1065-1078.

Yoshida Y, Kiyosue T, Katagiri T, Ueda H, Mizoguchi T, Yamaguchi-Shinozaki K, Wada K, Harada Y, Shinozaki K (1995) Correlation between the induction of a gene for Δ1-pyrroline-5-carboxylate synthetase and the accumulation. *Plant J.* 7: 751–760.

Vance CP, Kirk TK, Sherwood RT (1980) Lignification as a mechanism of disease resistance. *Annu. Rev. Phytopathol.* 18: 259-288.

Verbruggen N, Hua XJ, May M, Van Montagu M. (1996) Environmental and developmental signals modulate proline homeostasis: evidence for a negative transcriptional regulator. *Proc. Natl. Acad. Sci. USA* 93: 8787-8791.

Verslues PE, Kim YS, Zhu JK (2007) Altered ABA, proline and hydrogen peroxide in an Arabidopsis glutamate: glyoxylate aminotransferase mutant. *Plant Mol. Biol.* 64: 205-217.

In: Ammonia: Structure, Biosynthesis... ISBN: 978-1-62100-502-5
Editors: V.A. Fekete, et al, pp. 99-112

Chapter 5

ATMOSPHERIC CONCENTRATION OF AMMONIA, NITROGEN DIOXIDE, NITRIC ACID AND SULFUR DIOXIDE BY MEMBRANE-TYPE PASSIVE METHOD AND THEIR EMISSION INVENTORY IN JAPAN

Yoshinori Nishikawa* and Akiyoshi Kannari
Kitashinmachi, Ikoma City, 630-0245, Japan.

ABSTRACT

Annual emission map for NH_3, NO_X and SO_2 in Japan was shown according to the EAGrid2000-Japan emission database. The median emission of NH_3, NO_X and SO_2 in the 10 X 10 km grid was 0.37, 0.69, 0.078 ton/km^2/y, respectively, while that at the 30 sites was 2.4, 22, 3.3 ton/km^2/y, respectively. Monthly emission for NH_3 showed apparent seasonal trends, being high in summer and low in winter. In the case of NO_X and SO_2, the emission was slightly high in winter and low in summer and constant through the year, respectively.

Atmospheric concentration of NH_3, NO_2, HNO_3 and SO_2 by the passive method was compared with corresponding emission inventory. Average concentrations of HNO_3, SO_2, NH_3 and NO_2 were 5.6-39.7, 11-146, 34-175 and 93-1191 nmol/m^3, respectively. The emission inventory

* TEL/FAX: 81-743-20-7944, E-mail: yonishikawa@gaia.eonet.ne.jp

flux of SO_2, NO_X and NH_3 was investigated within $1km^2$, $100km^2$ and $1300km^2$ zones including the sampling sites. The correlation for NH_3 was significant in the three emission zones. The correlations for NO_2, HNO_3 and SO_2 were also significant, although with some exception. As the emission inventory included rather high stack (more than 25m) facilities combustion sources, the correlation probably was good in large sphere rather than the small sphere.

Monthly average concentration of NH_3, NO_2, HNO_3 and SO_2 was shown at the sites where performed relatively long term survey during FY2003-2006. Monthly concentration of NH_3 was high from July to November, while monthly emission of NH_3 was high from June to September (summer) and low from December to March (winter). The temporal trend of NO_2 concentration was high in winter and low in summer similar to that of NO_X emission. In contrast, the trend of HNO_3 concentration was high in summer and low in winter, reverse to that of NO_X emission. There was not particular seasonal trend of SO_2 concentration, where SO_2 emission had also not seasonal variation.

Keywords: emission inventory, passive method, NH_3, NO_X, SO_2.

INTRODUCTION

An atmospheric emission inventory, called "EAGrid2000-Japan" [1], is developed to provide reliable fundamental data in the year 2000. This inventory, emissions from the Japanese region, is estimated in detail with high temporal and spatial resolution: temporally, hourly by month, and spatially, 30" in latitude and 45" in longitude, equivalent to around 1km × 1km resolution.

On the other hand, passive sampling is a convenient method to determine the regional distribution of concentration at many spots because the method is inexpensive and easily adapted to different conditions. Field measurement of acidic gases such as SO_2, HCl, HNO_3, NO_2 and acidic potential gases such as O_3 and NH_3 are conducted with the passive method by the acid deposition survey network in Osaka Prefecture [2] and the Japan Environmental Laboratories Association (Acid Deposition Survey Sectional Meeting of Japan Environmental Laboratories Association [3-6].

Previous paper [7] described relationship between gas concentration determined by the passive method and corresponding gas emission inventory within Osaka Prefecture. This paper reports the same passive method carried out by Acid Deposition Survey of Japan Environmental Laboratories

Association (JELA) as well as Osaka Prefecture's Survey to investigate relation between of the passive method and the emission inventory.

METHODS

1. Emission Inventory

"EAGrid2000-Japan" is a database containing estimates of emissions per grid cell (1km X 1km at the most detailed resolution) of air pollutants such as SO_2, NO_X and NH_3. The whole area of Japan is divided into Small Square like a mesh based on latitudinal and longitudinal line called third-order mesh code. The emission inventory consists of large combustion sources (Power plant, Waste incineration and Other combustion) which classified three stack height (<25m, 25-100m and >100m) in each furnace, small combustion sources (Small combustion, Small incinerator and Field burning), moving sources (Road, Navigation, Aviation and Off road) and ammonia sources (Agriculture, Human and Pet and Other NH_3 sources).

Standard area mesh code [8] consists of first-order mesh code (80 X 80 km), second-order mesh code (10 X 10 km) and third-order mesh code (1 X 1 km). Emission flux is used third-order (1 km^2) and second-order (100 km^2) mesh code zone including passive sampling sites, and 13 blocks (1300 km^2) which is constituted nine brocks adjacent to the sampling sites and four direction (N, E, S and W) blocks around them to investigate relation between gas concentrations and corresponding emission inventories. The 13 blocks zone corresponds to within a 20-km radius of the sampling sites.

2. Passive Method

Field monitoring was carried out on a monthly basis at 30 sites from April 2003 to March 2006 in Acid Deposition Survey of JELA. Table 1 shows passive sampling site information such as latitude, longitude, altitude and third-order mesh code. Figure 1 shows geographical location of the passive sampling sites in the JELA survey.

Preparation of sampler, sampling and analytical procedure were performed according to literature [9-10].

Table 1. Passive sampling sites information

Site	Latitude (°N)	Longitude (°E)	Altitude (m)	3rd order meshcode	Local Government
1	43.0817	141.3336	12	64415208	Hokkaido
2	36.7547	139.5975	620	55390498	Tochigi
3	36.6006	139.9403	140	54397715	Tochigi
4	35.5256	140.0686	2	53402035	Chiba
5	35.0000	138.3900	10	52383391	Toyama
6	36.7006	137.0994	10	55370038	Shizuoka
7	36.0739	136.2617	11	54360280	Fukui
8	35.5706	136.6975	140	53362586	Gifu
9	34.7422	137.3842	20	52370380	Aichi
10	35.1006	136.9869	60	52365719	Nagoya
11	34.9539	136.6139	10	52363438	Mie
12	34.9667	135.9333	170	52353764	Shiga
13	34.9853	135.8986	87	52353782	Shiga
14	34.8700	135.7000	87	52352546	Shiga
15	34.8742	135.6936	70	52352545	Kyoto
16	34.6794	135.5350	2	52350413	Osaka
17	34.8250	135.4475	70	52351386	Osaka
18	34.6675	135.6389	9	51357591	Osaka
19	34.4864	135.5119	77	51355481	Osaka
20	34.6489	135.1317	15	51357170	Hyogo
21	34.6700	135.8200	94	52350605	Nara
22	34.2000	135.2094	15	51352136	Wakayama
23	35.3494	134.4953	820	53340410	Tottori
24	35.5197	134.2156	2	53342127	Tottori
25	35.4906	133.8881	2	53331781	Tottori
26	35.4400	133.8500	17	53331628	Tottori
27	35.4300	133.3500	7	53331218	Tottori
28	34.1531	131.4331	13	51311374	Yamaguchi
29	34.0694	134.5608	2	51340484	Tokushima
30	33.5861	133.8667	230	50333609	Kochi

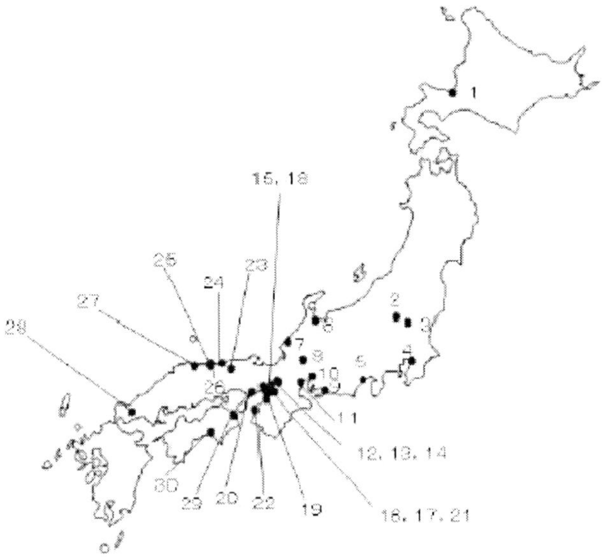

Figure 1. Geographical locations of the sampling sites.

RESULTS AND DICUSSION

1. Annual Emission Map

Figure 2 shows the second-order mesh distribution of annual emission flux for NH_3, NO_X and SO_2 in Japan according to the EAGrid2000-Japan emission database with the use of MANDARA GIS free software [11]. Maximum (Median) annual emission flux of NH_3, NO_X and SO_2 in the 10 X 10 km grid was 18 (0.37), 251 (0.69), 177 (0.078) ton/km^2/y, respectively, while that at the 30 sites was 9.9 (2.4), 133 (22), 86 (3.3) ton/km^2/y, respectively. Distribution of the 3 pollutants consistently showed high emission in the mega urban area such as around Tokyo, Nagoya, and Osaka, or in agricultural areas in the case of NH_3.

2. Monthly Emission Variation

While the inventory is estimated in detail with hourly by month for each source, monthly variations of 1 X 1 km mesh emissions for NH_3, NO_X and

SO_2 at the 30 sites are shown in Figure 3. The emission flux of NH_3 showed apparent seasonal trends, being high in summer and low in winter except for the Site 4.

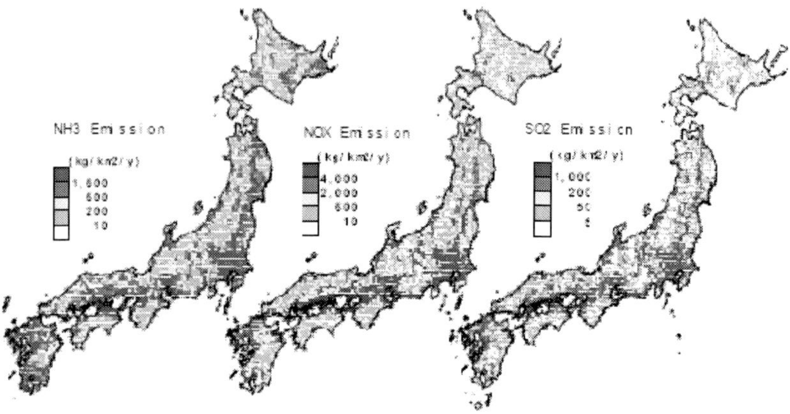

Figure 2. Annual emission map for NH_3, NO_X, and SO_2 using 10 X 10 km mesh data in Japan.

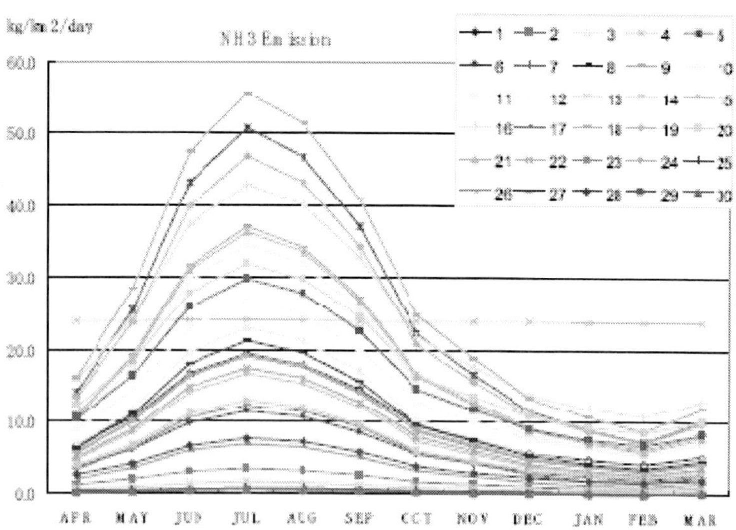

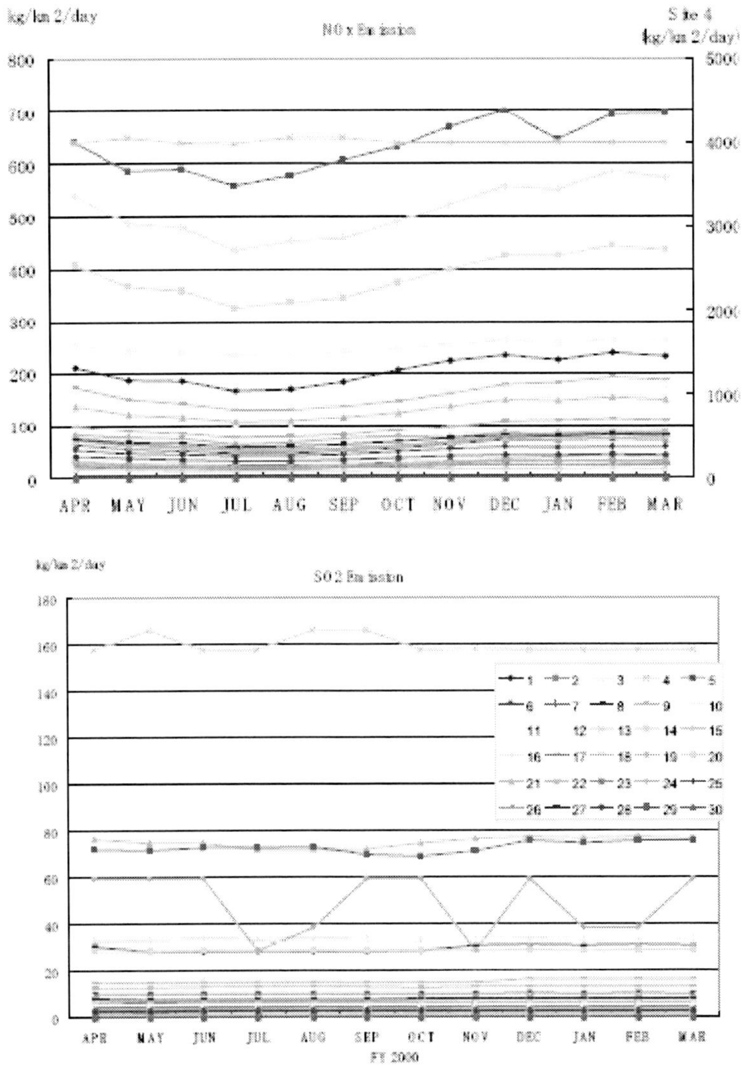

Figure 3. Monthly emission within 1 X 1 km mesh for NH_3, NO_X and SO_2 at the 30 sites.

In the Site 4, or mesh code 53402035, there were combustion facilities installed 25-100m and >100m height stacks (1.87 and 21.5 kg/km^2/day, respectively, through the year) as large combustion sources. The flux of NO_X was slightly high in winter and low in summer. The flux at the Site 4 was much higher comparing with the other sites. There was also combustion facilities installed more than 25m-height stacks (1161-2330 $kg/km^2/day$ through the year) at the Site 4. For SO_2 emission, flux was constant through the year for the most sites. In 53402035 (Site 4) and 53342127 (Site 24), there also had combustion facilities more than 25m height stacks (38-126 and 21-52 $kg/km^2/day$, respectively, through the year) as large combustion sources. It is predicted that high stuck effluents affect the wide range of grid, but have little impact on just neighboring grid. The emission flux of the three gases dropped sharply being revised to exclude high stack (25-100m and >100m) facilities' inventory.

3. Relation Atmospheric Gas Concentration with Its Emission Inventory

Atmospheric concentration of NH_3, NO_2, HNO_3 and SO_2 by the passive method is compared with corresponding emission inventory. Table 2 shows concentration of the gases at 30 sites during FY 2003-2006. Average concentrations of HNO_3, SO_2, NH_3 and NO_2 were 5.6-39.7, 11-146, 34-175 and 93-1191 $nmol/m^3$, respectively. Table 3 shows the emission inventory of SO_2, NO_X and NH_3 within $1km^2$, $100km^2$ and $1300km^2$ zone including the sampling spots. The emission flux of SO_2 was large difference among the three zone categories in the Sites 4, 12, 13, 21, 22, 24 and 29. That of NO_X was also large difference in the Sites 4, 16, 20 and 29. For NH_3, it was relatively small difference compared with SO_2 and NO_X due to different source features. Table 4 presents correlation between average concentration of the gases and their corresponding gases emission inventory. The correlation for NH_3 was significant in $1km^2$, $100km^2$ and $1300km^2$ emission areas. The concentration of NO_2 and HNO_3 was correlated with NO_X as a precursor. The correlations for NO2 were less than 10% significant level in the $1km^2$ and $100km^2$ zone, although 10% significant level in the $1300km^2$. The correlations for HNO_3 were 1% significant level in the $1km^2$ and $100km^2$ zone, although less than 10% significant level in the $1km^2$. Those for SO_2 were 5% significant level in the $1km^2$, less than 10% significant level in the $100km^2$ and 1% significant level in the $1300km^2$. As the emission inventory included rather high stack (more than 25m) facilities combustion sources, the correlation probably was good in large sphere rather than second-order or third-order mesh.

Table 2. Concentration of HNO_3, SO_2, NH_3 and NO_2 by the passive method during FY 2003-2006

Site	SO_2 Emission ($t/km^2/y$)			NO_x Emission ($t/km^2/y$)			NH_3 Emission ($t/km^2/y$)		
	$1km^2$	$100km^2$	$1300km^2$	$1km^2$	$100km^2$	$1300km^2$	$1km^2$	$100km^2$	$1300km^2$
1	10.8	8.9	3.7	75.1	35.7	9.6	3.7	2.4	1.6
2	2.3	0.5	0.3	10.5	7.2	1.1	0.7	0.2	0.4
3	5.6	2.2	2.6	23.8	13.1	6.8	4.1	2.2	2.7
4	58.4	35.7	15.5	1465.1	117.0	46.1	8.7	8.5	3.1
5	3.6	7.6	3.6	21.9	30.9	6.7	8.9	3.5	0.9
6	1.1	4.1	6.6	13.4	21.1	12.2	2.1	1.8	1.6
7	1.5	0.6	1.7	7.8	6.2	3.5	2.2	0.6	0.8
8	0.0	0.3	3.0	0.0	1.3	4.4	0.1	2.4	1.6
9	5.6	2.9	2.2	34.8	19.6	7.4	6.7	9.9	4.4
10	3.0	11.0	9.4	43.2	97.5	37.8	5.1	5.2	4.3
11	11.9	3.1	8.1	90.8	23.7	19.4	6.1	3.9	2.1
12	2.1	23.9	3.8	53.1	40.7	16.7	3.2	1.5	1.4
13	0.6	23.9	6.1	5.7	40.7	10.8	0.3	1.5	2.2
14	2.7	3.6	3.0	23.8	39.8	5.8	2.4	2.6	2.5
15	0.5	3.6	2.9	10.8	39.8	19.3	2.9	2.6	4.0
16	12.3	15.4	12.2	185.6	133.0	49.9	8.6	6.4	7.0
17	1.5	5.3	10.8	24.3	76.1	40.9	3.4	4.6	5.1
18	4.7	1.2	7.6	53.5	12.8	31.5	10.0	3.4	5.3
19	1.1	2.2	3.6	24.8	22.0	20.5	8.3	4.0	4.6
20	10.5	11.9	11.5	141.2	42.6	25.2	6.5	0.7	2.1
21	27.5	3.0	3.3	43.2	16.3	16.2	6.5	2.8	3.5
22	1.0	77.9	0.2	9.5	88.1	0.6	3.1	3.7	0.4
23	0.0	0.1	0.1	0.0	0.7	0.6	0.1	0.3	0.6
24	17.9	8.9	0.9	23.1	29.8	2.6	3.5	3.6	0.6
25	0.0	1.1	0.9	1.2	5.4	2.3	0.2	1.0	1.4
26	0.8	0.1	0.6	3.2	0.9	1.9	1.2	0.5	1.8
27	2.9	1.1	2.2	25.9	6.9	4.0	3.8	2.4	1.4
28	0.9	7.1	2.5	14.1	18.9	8.8	1.5	1.7	0.9
29	26.6	1.3	3.5	231.3	9.6	9.0	5.8	1.1	2.9
30	0.0	0.0	0.2	0.0	0.5	1.0	0.1	0.2	0.4

Table 3. Emission inventories of SO_2, NO_X and NH_3 for three zones including the sampling sites

Site	HNO_3 (nmol/m^3)		SO_2 (nmol/m^3)		NH_3 (nmol/m^3)		NO_2 (nmol/m^3)	
	AVE ± SD	N	AVE ± SD	N	AVE ± SD	N	AVE ± SD	N
1	7.4 ± 5.3	30	---------		---------		---------	
2	6.5 ± 8.1	33	24 ± 13	32	34 ± 19	32	513 ± 92	32
3	21.9 ± 20.8	33	75 ± 26	32	173 ± 36	32	856 ± 178	32
4	22.0 ± 12.5	36	---------		---------		---------	
5	39.7 ± 32.2	12	47 ± 17	12	61 ± 8	12	819 ± 197	12
6	20.3 ± 18.0	48	40 ± 13	40	86 ± 58	40	---------	
7	16.2 ± 14.3	11	59 ± 31	11	---------		---------	
8	5.7 ± 3.5	12	11 ± 6	12	---------		---------	
9	24.5 ± 16.8	48	79 ± 21	48	---------		---------	
10	27.4 ± 21.3	36	91 ± 23	36	158 ± 23	36	776 ± 302	36
11	23.0 ± 17.7	24	146 ± 45	24	131 ± 31	24	866 ± 349	21
12	17.8 ± 11.7	12	67 ± 17	12	---------		---------	
13	20.5 ± 14.6	24	40 ± 18	24	---------		960 ± 489	9
14	37.0 ± 27.0	22	56 ± 19	22	55 ± 24	22	706 ± 377	22
15	12.6 ± 10.3	45	37 ± 17	45	109 ± 38	33	808 ± 253	32
16	37.5 ± 30.7	48	104 ± 29	48	175 ± 39	24	977 ± 431	48
17	26.1 ± 18.5	36	50 ± 17	36	64 ± 22	12	509 ± 226	36
18	31.2 ± 23.9	36	64 ± 20	36	154 ± 44	12	762 ± 386	36
19	28.0 ± 19.5	30	50 ± 16	27	94 ± 28	12	533 ± 225	28
20	30.9 ± 22.1	36	123 ± 39	35	95 ± 39	35	1191 ± 527	24
21	24.8 ± 19.0	27	36 ± 23	24	97 ± 28	27	---------	
22	21.5 ± 14.4	48	64 ± 22	48	52 ± 13	24	596 ± 247	21
23	9.2 ± 6.8	45	23 ± 15	33	---------		---------	
24	13.5 ± 10.6	45	69 ± 38	33	---------		---------	
25	7.6 ± 4.9	46	23 ± 10	32	---------		---------	
26	13.9 ± 8.8	10	49 ± 19	10	---------		---------	
27	13.2 ± 7.4	10	65 ± 28	10	---------		---------	
28	17.3 ± 14.6	34	32 ± 12	34	49 ± 11	34	441 ± 210	31
29	24.2 ± 17.0	36	83 ± 27	36	103 ± 35	35	686 ± 339	35
30	5.6 ± 2.7	41	20 ± 17	40	106 ± 31	41	93 ± 41	41

4. Monthly Variation of Gas Concentrations

Figure 4 shows monthly average concentration of NH_3, NO_2, HNO_3 and SO_2 at the Sites 2, 6, 10, 16, 22, 28, 29 and 30 where were conducted relatively long term survey sites during FY2003-2006. Monthly concentration of NH_3 was high from July to November, while monthly emission of NH_3 was high from June to September (summer) and low from December to March (winter) as described previously. The temporal trend of NO_2 concentration was high in winter and low in summer similar to that of NO_X emission. On the other hand, the trend of HNO_3 concentration was high in summer and low in winter, reverse to that of NO_X emission. The difference of concentration at the sites was little in winter compare with the other season.

Table 4. Correlation between average concentrations and their emission inventories

Concentration Gases (Y)	Emission Gases (X)	Emission Area	Regression Line	r	N
		1km^2	Y=7.0X+66	0.473^{b}	
NH_3	NH_3	100km^2	Y=11X+70	0.414^{c}	18
		1300km^2	Y=14X+60	0.619^{a}	
		1km^2	Y=1.7X+620	0.450^{c}	
NO_2	NO_X	100km^2	-------	0.328	17
		1300km^2	-------	0.380	
		1km^2	-------	0.124	
HNO_3	NO_X	100km^2	Y=0.13X+16	0.478^{a}	30
		1300km^2	Y=0.36X+15	0.535^{a}	
		1km^2	Y=1.8X+49	0.423^{b}	
SO_2	SO_2	100km^2	-------	0.154	28
		1300km^2	Y=5.2X+37	0.597^{a}	

Y: nmol/m^3, X: kg/km^2/year, r: correlation coefficent, N: number of data sets

Significance levels are denoted as: ^{a}: $p < 0.01$, ^{b}: $p < 0.05$, ^{c}: $p < 0.1$

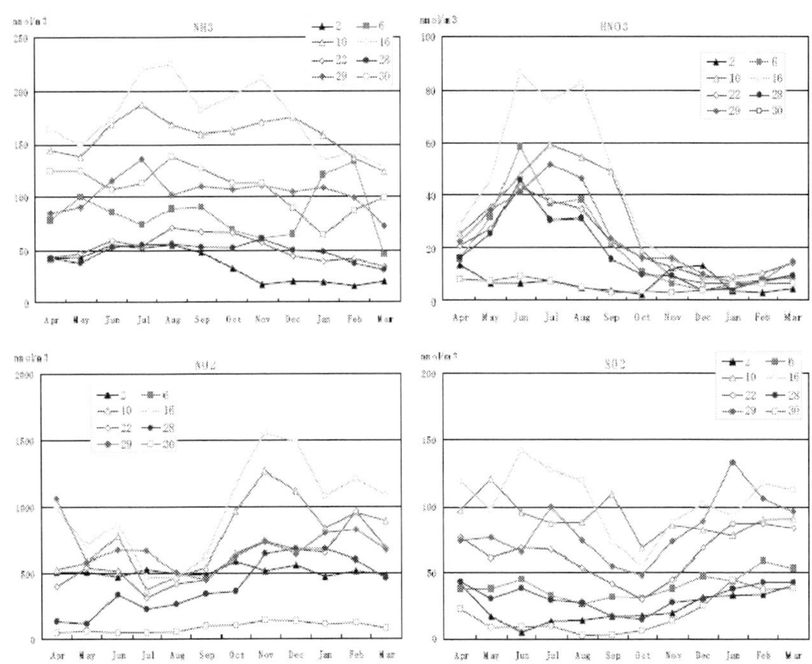

Figure 4. Monthly variation of atmospheric gas concentration during FY 2003-2006.

It was suggested that HNO_3 concentration depended on environment of oxidative reactivity rather than NO_X emission itself in the atmosphere. There was not particular seasonal trend of SO_2 concentration, where SO_2 emission had also not seasonal variation but was constant.

CONCLUSIONS

It was compared that atmospheric concentration of NH_3, NO_2, HNO_3 and SO_2 by the passive method during FY 2003-2006 at 30 sites was compared with corresponding gases such as NH_3, NO_X and SO_2 emission inventory. The correlation for NH_3 was significant in the three zones emission including the sampling spots. The correlation NO_2 concentration and NO_X emission was significant in the 1 X 1 km mesh, but was not significant in 10 X 10 km mesh and 1300 km^2 zone. The correlations for HNO_3 and SO_2 were 1% significant level in the 1300km^2. As the emission inventory included rather high stack (more than 25m) facilities combustion sources, the correlation probably was good in large sphere rather than second-order (100km^2) or third-order (1km^2) mesh.

Monthly concentration of NH_3 was high from July to November, while monthly emission of NH_3 was high from June to September (summer) and low from December to March (winter). The temporal trend of NO_2 concentration was high in winter and low in summer similar to that of NO_X emission. On the other hand, the trend of HNO_3 concentration was high in summer and low in winter, reverse to that of NO_X emission. It was suggested that HNO_3 concentration depended on environment of oxidative reactivity rather than NO_X emission itself in the atmosphere. There was not particular seasonal trend of SO_2 concentration, where SO_2 emission had also not seasonal variation but was constant.

ACKNOWLEDGEMENTS

The authors express their appreciation to the local government environmental institute and researchers who participated in the passive method monitoring in the Japan Environmental Laboratories Association survey.

REFERENCES

[1] Kannari, A., Yutaka Tonooka, Y., Baba, T. and Murano, K.: Development of multiple-species 1km X 1km resolution hourly basis emissions inventory for Japan. *Atmospheric Environment*, 41, 3428-3439 (2007).

[2] Nishikawa, Y. and Hatanaka, H. (2008): Measurement of Acidic Gases within Osaka Prefecture using a Simple Passive Sampling Method. *Bull. Res. Inst. Env. Agr. Fish. Osaka*, 1, 6-12, (2008), (in Japanese with English abstract).

[3] Acid Deposition Survey Sectional Meeting of Japan Environmental Laboratories Association: The report of acid deposition survey in Japan, phase 4. *Journal of Japan Environmental Laboratories Association* 30, 58-135, (2005), (in Japanese).

[4] Acid Deposition Survey Sectional Meeting of Japan Environmental Laboratories Association: The report of acid deposition survey in Japan, phase 4. *Journal of Japan Environmental Laboratories Association* 31, 118-186, (2006), (in Japanese).

[5] Acid Deposition Survey Sectional Meeting of Japan Environmental Laboratories Association: The report of acid deposition survey in Japan, phase 4. *Journal of Japan Environmental Laboratories Association* 32, 78-152, (2007), (in Japanese).

[6] Acid Deposition Survey Sectional Meeting of Japan Environmental Laboratories Association: The report of acid deposition survey in Japan, phase 4. *Journal of Japan Environmental Laboratories Association* 33, 126-196, (2008), (in Japanese).

[7] Nishikawa, Y. and Kanari, A.: Atmospheric Concentration of Ammonia, Nitrogen Dioxide, Nitric Acid and Sulfur Dioxide by Passive Method within Osaka Prefecture and Their Emission Inventory. *Water Air Soil Pollut.*, in press (2010), (DOI 10.1007/s11270-010-0472-3).

[8] http://homepage2.nifty.com/spmap/doc/etc/MeshDefine.html.

[9] Nishikawa, Y., Murano, K. and Mukai, H.: Comparison of Sampling Resistance for One to Three Sheets of Membrane Type Passive Sampler. *Water Air Soil Pollut.*, 197, 241-247, (2009).

[10] Tokai-Kinki-Hokuriku Branch of Japan Environmental Laboratories Association: Examination for practical use of simple passive sampling methods --- Collaborative investigation by Tokai-Kinki-Hokuriku Branch of Environmental Laboratories Association in Japan ---. *Journal*

of Japan Environmental Laboratories Association 29, 25-35, (2004), (in Japanese with English abstract).
[11] http://ktgis.net/mandara/index.php.

In: Ammonia: Structure, Biosynthesis... ISBN: 978-1-62100-502-5
Editors: V.A. Fekete, et al, pp. 113-126 © 2012 Nova Science Publishers, Inc.

Chapter 6

CONCENTRATION GRADIENT MEASUREMENTS AND FLUX CALCULATION OF ATMOSPHERIC AMMONIA OVER GRASSLAND (BUGAC-PUSZTA, HUNGARY)

T. Weidinger*[*,1], A. Pogány[2], L. Horváth[3], A. Machon[1,4], Z. Bozóki[5], Á. Mohácsi[5], K. Pintér[4], Z. Nagy[4], A. Z. Gyöngyösi[1], Z. Istenes[1] and Á. Bordás[1]

[1] Eötvös Loránd University, Budapest, Hungary.
[2] University of Szeged, Szeged, Hungary.
[3] Hungarian Meteorological Service, Budapest, Hungary.
[4] Szent István University, Gödöllő, Hungary.
[5] Hungarian Academy of Sciences, Szeged, Hungary.

ABSTRACT

Ammonia flux has been monitored continuously since July 2008 over semi-natural grassland at the Hungarian NitroEurope site '*Bugac-puszta*' on the Great Hungarian Plain. Results presented here are based on the data obtained from July to September, i.e., during the vegetation period. The instrument used for ammonia concentration gradient

* Corresponding author: E-mail address: weidi@ludens.elte.hu, Phone: +36 1 3722500/ext. 6612, Fax: +36 1 3722904

measurement was a novel diode laser based photoacoustic device combined with preconcentration sampling (WaSul-Flux), developed at the University of Szeged. Ammonia concentration measurements were performed at three different levels (0.5 m, 1.3 m and 3 m), on a cc. 30-minute accumulation interval. The three inlets were moved automatically to the same level (1.3 m) twice a week by a remote controlled automated system to check the precision of the measurement.

The turbulent flux of ammonia was calculated using the similarity theory based on eddy covariance data of momentum, heat, water vapor and carbon dioxide fluxes (provided by a CSAT3 sonic anemometer and a LICOR-7500 open path CO_2/H_2O sensor), in view of the friction velocity (u_*) and the Monin-Obukhov length scale (L). Sensitivity analyses of ammonia flux calculation as (i) calculation of ammonia gradient, (ii) choice of universal function and (iii) application of different gradient and profile techniques, have been investigated.

The diurnal variation of the ammonia concentration and flux has also been investigated. During the studied period the net daytime emission and nocturnal deposition were observed with large deviation exceeding the average flux values both during day and night.

The daily mean ammonia concentrations were compared to data measured at the Hungarian background air quality monitoring station (K-puszta) ~20 km far from the Bugac-puszta site, and fairly good agreement was found between the two datasets.

1. INTRODUCTION

The NitroEurope EU Framework 6th Integrated Project aims at a detailed investigation of the biosphere-atmosphere exchange of different nitrogen (N) compounds. The objectives of this international project are (i) to establish robust datasets of N fluxes and net greenhouse-gas exchange (NGE) as a basis to investigate interactions and assess long-term change, (ii) to quantify the effects of past and present global changes (climate, atmospheric composition, land-use/land-management) on C-N cycling and NGE, (iii) to simulate the observed fluxes through refinement of plot-scale models, and (iv) to scale up N and NGE fluxes for terrestrial ecosystems to regional and European levels [1-3]. One of the measurement stations of the NitroEurope network has been established in Centeral Hungary (*Bugac-puszta*). Four Hungarian institutes are involved in the project: the Forest Research Institute (with an observing system of N fluxes and pools), the Photoacoustic Research Group at the University of Szeged (with the development of photoacoustic ammonia monitoring instruments), the Institute of Botany and Ecophysiology at Szent

István University (with a network of manipulation experiments and micrometeorological measurements) and the Department of Meteorology at Eötvös Loránd University (with an observing system of N fluxes and pools, micrometeorology and flux calculations).

The N balance of terrestrial ecosystems is mainly determined by surface-atmosphere exchange processes. This exchange is bi-directional: wet and dry deposition from the atmosphere is a major N input to ecosystems (especially in case of non-treated sites without fertilizer or manure application), while at the same time a significant amount of N compounds is emitted by the biosphere. The net flux is the sum of deposition and emission. Ammonia is emitted mainly by plants through the stomata, while other nitrogen compounds (NO, N_2O and N_2) are emitted dominantly by the soil. Soil emission of ammonia can only be observed above alkaline soil (pH > 7).

The timescale of these exchange processes ranges from some hours to several years. To understand the small timescale processes (e.g., plant physiology) on-line ammonia flux measurements (with high time resolution and accuracy) are needed instead of long-term, time averaged flux calculations.

The aim of this chapter is to present the research and technological development to estimate net NH_3 flux between the atmosphere and a grassland surface in Central Hungary (*Bugac-puszta*) using a novel method (the combination of diode laser based photoacoustic spectroscopy and preconcentration sampling).

2. Measurement Site and Instrumentation, Dataset and Quality Control

The *Bugac-puszta* site is located on semi natural, semi arid, sandy grassland in the Kiskunság National Park (46°41′ N, 19°36′ E) in the Great Hungarian Plain. The dominant plant species are *Festuca pseudovina*, *Carex stenophylla*, *Salvia pratensis* and *Cynodon dactylon*. The flora is sensitive to external effects, but Hungarian Grey Cattle (*Bos p. hungaricus*) has been grazing for centuries in equilibrium with the grass ecosystem (0.5–0.8 cattle ha^{-1} in the periods of investigation). The landscape is flat, the average altitude is 110 m AMSL. The soil is a Chernozem-like sandy soil (with a mean composition of sand : silt : clay at a ratio of 87 : 8 : 5, in the upper 30 cm layer of the soil). The climate is continental: the annual mean temperature is 10.5 °C and the average precipitation is about 530 mm $year^{-1}$ [4].

Ammonia flux has been monitored continuously since July 2008, while the measurement of basic micrometeorological parameters, radiation, energy budget components and CO_2, N_2O fluxes started in 2002 (EU 5th Greengrass program) [5]. Determination of nitrogen fluxes and estimation of N balance has been performed since 2006 (EU 6th NitroEurope program, Figure 1). Continuous NO, NO_x and O_3 profile measurements and chamber measurements of NO, N_2O, CH_4 have also been carried out [6].

The bi-directional flux of NH_3 between the atmosphere and the biosphere (including soil emission) has been determined using the photoacoustic method. The photoacoustic effect is based on the absorption of modulated laser light in a photoacoustic cell, which creates an acoustic wave through non-radiative relaxation of the excited molecules. The amplitude of the acoustic wave, which is sampled with a sensitive microphone, is proportional to the concentration of the absorbing component [7-9].

Advantages of this system include: (i) linear response over more than four orders of magnitude, high selectivity (i.e., insensitivity to the presence of other components), relatively short response time (15–45 minutes), high sensitivity (detection limit below 1 ppb) and accuracy in the few percentage range, (ii) simple construction and capability of long-term automatic operation. The performance of the instrument proves that is has a potential to become an alternative of currently used ammonia flux monitoring instruments [10-12].

Ammonia measurements have been performed at three different heights (0.5 m, 1.3 m and 3 m) above canopy level, with a cc. half hour averaging interval. The inlets are moved automatically to the same height twice a week for cross-calibration, and concentration values measured at different heights are corrected according to the results of cross-calibration. The uncertainty of concentration measurement may cause significant uncertainty in flux calculation only in case of low concentrations or concentration gradients. The turbulent flux of ammonia is calculated using the similarity theory based on eddy covariance fluxes of momentum, heat, water vapor and carbon dioxide (measured by a CSAT3 sonic anemometer and LICOR-7500 open path CO_2/H_2O sensor).

Flux calculation and quality control was performed using the methods of the EU 6th CarboEurope program [5, 13], which means omitting data during nocturnal highly stable conditions, during fog or dew formation and during instationarity.

The friction velocity (u_*) and the Monin-Obukhov length scale (L) were determined from half hourly momentum (τ) and sensible heat flux (H) data.

Figure 1. The Bugac-puszta station in the Great Hungarian Plain with the eddy covariance measurement system (CSAT-3 sonic anemometer and LICOR-7500 open path CO_2/H_2O sensor), inlets for NO, NO_x, O_3 profile measurements, radiation (global, reflected, net, PAR) measurement pole (in background) and dynamic chambers for NO flux calculation (in foreground).

In cases when momentum and sensible heat flux data were not available (due to the measurement error of the latent heat or CO_2 flux) for a certain half hour period, they were calculated using regression relationships from raw fluxes, available energy ($A = Rn - G$) and measured wind speed data. The Schotanus correction [14, 15] was performed in all cases for the correction of raw sensible heat fluxes. The friction velocity (u_*) was estimated from the measured wind speed.

3. Ammonia Flux Calculation

The ammonia flux was calculated on the base of the similarity theory using the profile method. Based on model calculations, the general form of the universal functions [11, 16] has been chosen in our study and according to Ref. [17] a constant value has been considered in case of highly stable conditions. Several publication report ammonia flux calculations performed with the application of different universal functions [18]. The results are highly similar, which means that the uncertainty of flux calculation is determined mainly by the uncertainty of gradient measurements [19].

Let us consider the main steps of the flux calculation. The turbulent fluxes (covariance) for momentum (τ), sensible heat (H) and ammonia (F_c), respectively, are described by the following equations:

$$\tau = -\rho_m \overline{w'u'} = \rho_m u_*^2 \,, \quad H = c_{pm}\rho_m \overline{w'T'} = -c_{pm}\rho_m u_* T_* \,,$$

$$F_c = \overline{w'c'} = -u_* c_* \,, \tag{1}$$

where ρ_m is the density of moist air, c_{pm} is the specific heat capacity of moist air at constant pressure, T', c', u' and w' are fluctuations of temperature, concentration, horizontal wind speed and vertical wind speed, respectively, u_*, T_*, and c_* denote the friction velocity, the dynamic temperature and dynamic concentration, respectively. The similarity theory provides a flux-profile relationship for the ammonia concentration profile:

$$\frac{\partial c}{\partial z} = \frac{c_*}{\kappa z}\varphi_{Fc}(\zeta) \,, \tag{2}$$

where $\zeta = z/L$ is the non-dimensional stability parameter, $\varphi_{Fc}(\zeta)$ represents the universal function for ammonia flux and z denotes the vertical coordinate. The Monin-Obukhov length scale (L) was determined as:

$$L = -\frac{(\tau/\rho_m)^{3/2}}{\kappa\beta(H/c_{pm}\rho_m)} = \frac{u_*^2}{\kappa\beta T_*} \,, \tag{3}$$

where $\boldsymbol{\kappa}$ **is the von Kármán constant,** $\beta = g/T$ is the stability parameter and g is the acceleration due to gravity. The process of trace gas (or moisture) transport is similar to the transport of sensible heat, which means that c_* is a passive characteristic of the turbulent motion ($\varphi_H(\zeta) = \varphi_{Fc}(\zeta)$).

The value of the universal function during neutral stratification ($\zeta \approx 0$) is 1 and a logarithmic profile approximation can be used. During stable cases ($\zeta > 0$) the log-linear profile approximation is applied, while during highly stable stratification ($\zeta \geq 1.5$) constant values are used:

$$\varphi_{F_c}(\zeta) = \varphi_H(\zeta) = 1 + 5\zeta, \ \zeta \leq 1.5,$$
$$\varphi_{F_c}(\zeta) = \varphi_H(\zeta) = 7.5, \quad \zeta \geq 1.5. \tag{4}$$

In unstable cases ($\zeta < 0$) the power law approximation is used:

$$\varphi_{F_c}(\zeta) = \varphi_H(\zeta) = (1 - 16\zeta)^{-1/2}. \tag{5}$$

The ammonia concentration profile is calculated with the integration of equation (2) as:

$$c(z_i) - c(z_j) = \frac{c_{*(i,j)}}{\kappa}[\ln(z_i / z_j) + \Psi_{F_c}(\zeta_i) - \Psi_{F_c}(\zeta_j)], \tag{6}$$

where $z_i > z_j$ (i, j =1, 2, 3) are the measurement heights above canopy level.

$$\Psi_{F_c}(\zeta_i) = \int (\varphi_{F_c}(\zeta) - 1) d \ln \zeta, \tag{7}$$

is the integral form of the universal function [17, 20]. **The lower boundary of the constant flux layer model is the roughness length height (z_0) with the assumption of zero wind speed at that level. The dynamic concentration (c_*) has been calculated from concentration differences between the sub-layers (8) with the least-square method as:**

$$\min \left(\sum_{\substack{i,j \\ i>j}} \left((c(z_i) - c(z_j)) - \frac{c_*}{\kappa}[\ln(z_i / z_j) + \Psi_{F_c}(\zeta_i) - \Psi_{F_c}(\zeta_j)] \right)^2 \right)^{\frac{1}{2}}. \tag{8}$$

Those cases were accepted when the (8) square errors of the differences in the measured and calculated concentrations did not exceed 0.75 ppb in all sub-layers.

4. Results and Discussions

Results of ammonia gradient measurements between 2st July and 5th October 2008 are summarized in this chapter. 1160 measurements have been made during 96 days, covering about 25% of the studied period and 30% of these data were filtered out during the quality control procedure.

Let us consider the daily mean concentration values first. They have been compared to the continental background air quality monitoring station at K-puszta (~20 km away from Bugac-puszta), where the so-called "three stage filter pack method" (applied in the EMEP network) was used for daily sampling of air, and the samples were analyzed spectrophotometrically (using the indophenol-blue method for ammonia). One of the longest ammonia concentration data series in Europe have been recorded at K-puszta [21, 22]. Data series from the two sites are showing relatively good agreement (r = 0.38). Higher concentration at *Bugac-puszta* site may be due to the effect of a farm nearby performing extensive breeding of grey cattle (Figure 2).

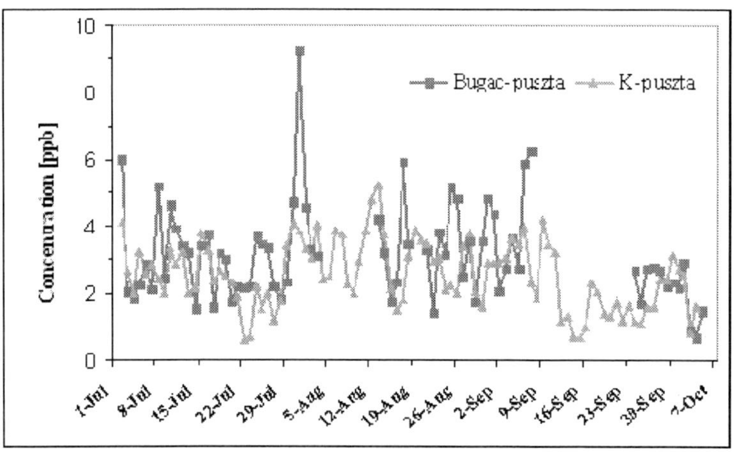

Figure 2. Comparisons of daily mean ammonia concentration data measured using photoacoustic (Bugac-puszta) and filter pack – spectrophotometry (K-puszta) methods at the two measurement sites for a 96-day period (between 2st July and 5th October 2008).

The results for July 2008 are shown below. Half hourly data of the friction velocity (u_*) and the sensible heat flux (H) are presented in Figure 3. Flux calculation and quality control methodology (also used in the

EU 6th CarboEurope program) have been applied [5, 13] as described in the previous section, while in about 20% of the period (especially during nighttime) gap filling was necessary to fill out missing or inappropriate data using raw fluxes, wind measurement and available energy data.

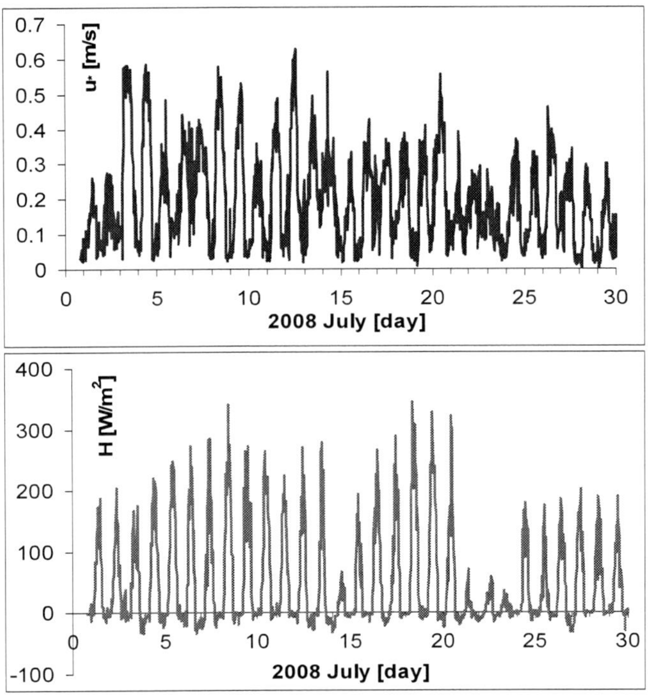

Figure 3. Half hourly values of dynamic velocity (u*) and sensible heat flux (H) at *Bugac puszta* in July, 2008.

There is a strong correlation ($r = 0.89$) between wind speed (u) and friction velocity (u_*). The latter is about the tenth (0.08) of wind speed measured at a height of 4 m. According to the daily variations of wind speed, higher values are obtained during the day and lower during the night. This effect has been amplified by strong nocturnal stability – weak mechanical turbulence. The hot dry days at the beginning of July are clearly visible in the course of the sensible heat flux (maximum values exceeding 250 W m^{-2}) together with the cloudy, humid days with low maxima.

Half hourly concentration values measured at each level and ammonia flux calculated using the similarity theory are depicted in Figure 4. The typical daily variation of ammonia concentration is demonstrated: higher values were observed during the day and lower values during nighttime. High concentration episodes showing the effect of advection (e.g. the effect of grazing in the vicinity of the station or the effect of the farm) are also remarkable.

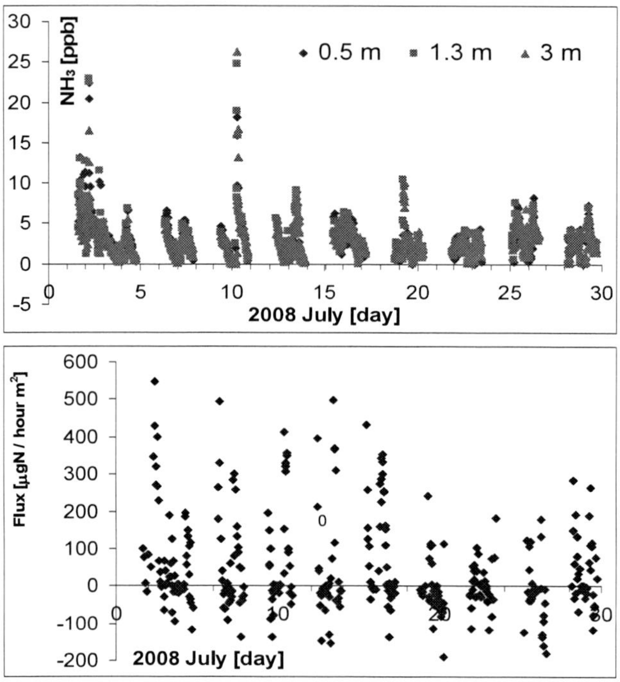

Figure 4. Measured ammonia concentration (top) at three levels (0.5 m, 1.3 m, 3 m) and the calculated fluxes (bottom) – *Bugac-puszta* in July, 2008.

During daytime, the measured concentration was decreasing with height (emission) and inverse profile (deposition) is typical during the night, showing the influences of the stomata. The calculated fluxes were higher during the first, warmer and drier period, while they were lower during the second, cooler and more humid period. The overall ammonia flux was positive: total emission exceeds total deposition. Similar results were obtained from recent measurements at Hortobágy (the largest „puszta" or grassland in the Great Hungarian Plain) for the summer months [12].

We note that the simplest model for the estimation of ammonia deposition uses average deposition velocities. The standard value used for estimating ammonia fluxes above Hungarian grass canopy is 0.3 cm s^{-1} [23], and is positive, which would result in deposition instead of emission. This deviation of the theory from the measurement is significant, and is most probably caused by an elevated emission of ammonia at the studied site due to the nearby cattle farm.

The average diurnal variation of the ammonia flux during the whole campaign (from 1st July to 4th October 2008, which is the second half of the vegetative period) has also been calculated. Results are presented in Figure 5. Since each half hour consists of 10 to 25 data that represent less than 20% of the total period, therefore the presented results are only rough estimates. Like in July, ammonia is emitted during the day and deposited during the night, however, the absolute values of the fluxes are relatively small. Large spreads are showing big uncertainty of the estimates: the values of the spreads are exceeding the averages of fluxes both during day and night.

The mechanism of the biosphere-atmosphere exchange of ammonia is described by the canopy compensation point model [11]. The canopy compensation point concentration, (i.e., the atmospheric concentration of ammonia above which deposition and under which emission occurs) is determined by a couple of the cuticular and stomatal features [12]. The N content of plant tissues is strongly dependent on the N content of the soil. Elevated N content in the soil enhances N uptake by the roots, which induces higher compensation point, as a result of which emission is preferred.

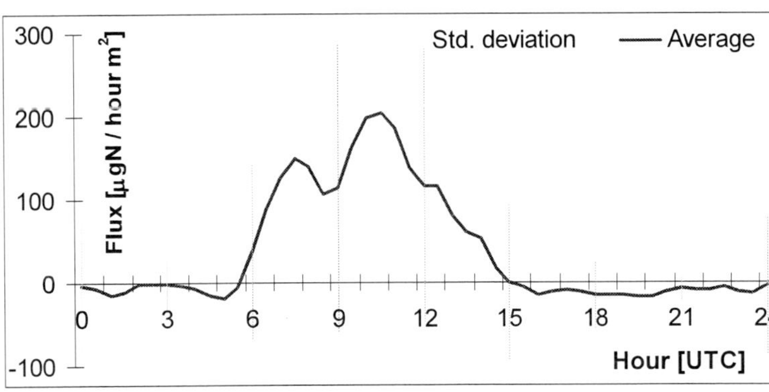

Figure 5. Average daily course of the ammonia flux during the measured period at Bugac-puszta site for a 96-day period (between 1st July and 4th October 2008).

Alkaline intracellular pH is also favorable for emission. During nighttime, when the stomata are closed, deposition dominates (see in Figure 5). During daytime, ammonia emission can occur, especially in the vegetation period (as it was observed using this study), however, over semi-natural grasslands net deposition is expected for a whole year [23].

Generally, yearly ammonia deposition is dominating on fields where there is no grazing or fertilization, while ammonia emission is observed in N loaded areas. Daily ammonia flux depends on soil conditions (available inorganic N), canopy compensation point concentration [12] and the studied period (vegetation or dormant season).

CONCLUSION

Ammonia concentration and flux measurements play the key role in the better understanding of the N-cycle and for the optimization of agricultural technologies. A micrometeorological measurement setup has been developed for continuous or campaign measurements of ammonia concentration gradient.

The recently developed photoacoustic ammonia gradient monitoring instrument is suitable for continuous monitoring of the bi-directional flux of ammonia. However, at low concentrations and advective situations, the assessment of the concentration gradient is problematic. During the application of our quality control procedure, 30% of the profile measurements had to be filtered out.

During the studied three months overall net ammonia emission has been observed, which implies the reconsideration of ammonia deposition estimates, based on a mean annual positive deposition velocity (+0.3 cm s^{-1}). The assessment of the ammonia budget, its spatial extrapolation, and the determination of the accuracy of the newly developed ammonia flux measuring system are the challenges for the near future.

ACKNOWLEDGEMENTS

This work has been supported by the NitroEurope Framework EU 6th IP and the GVOP 6-028-2005 project. The authors are grateful to the Kiskunság National Park for giving the opportunity to perform the measurements in the area of the park.

REFERENCES

[1] *http://www.nitroeurope.eu.*

[2] M.A. Sutton, E. Nemitz, J.W. Erisman, C. Beier, K. Butterbach Bahl, P. Cellier, W. de Vries, F. Cotrufo, U. Skiba, C. Di Marco, S. Jones, P. Laville, J.F. Soussana, B. Loubet, M. Twigg, D. Famulari, J. Whitehead, M.W. Gallagher, A. Neftel, C.R. Flechard, B. Herrmann, P.L. Calanca, J.K. Schjoerring, U. Daemmgen, L. Horvath, Y.S. Tang, B.A. Emmett, A. Tietema, J. Peñuelas, M. Kesik, N. Brueggemann, K. Pilegaard, T. Vesala, C.L. Campbell, J.E. Olesen, U. Dragosits, M.R. Theobald, P. Levy, DS.C. Mobbs, R. Milne, N. Viovy, N. Vuichard, J.U. Smith, P. Smith, P. Bergamaschi, D. Fowler and S. Reis, *Environ. Pollut.* 150, 1, 125–139 (2007).

[3] *Annual European Community greenhouse gas inventory 1990-2007 and inventory report* 2009. Submission to the UNFCCC Secretariat, 2009, European Environment Agency, EEA Technical report, No 4/2008, ISSN 1725–2237.

[4] L. Horváth, B. Grosz, A. Machon, J. Balogh, K. Pintér and Sz. Czóbel, *Community Ecology* 9 (Suppl.) 75–80 (2008).

[5] Z. Nagy, K. Pintér, Sz. Czóbel, J. Balogh, L. Horváth, Sz. Fóti, Z. Barcza, T. Weidinger, Zs. Csintalan, N.Q. Dinh, B. Grosz and Z. Tuba, *Agric. Ecosyst. Environ.* **121**, 21–29 (2007).

[6] T. Weidinger, L. Horváth, A. Machon, K. Pintér, Z. Barcza, A.Z. Gyöngyösi, Z. Nagy and Z. Tuba, 2008: *Uncertainties in the calculation of NO–NO_x–O_3 fluxes by the gradient and the profile methods,* Open Science Conference Reactive Nitrogen and the European Greenhouse Gas Balance. February 20th and 21st, 2008, Het Pand, Ghent, Belgium, Edited By the NitroEurope IP Secretariat, pp. 53–54.

[7] **H.** Huszár, **A.** Pogány, **Z.** Bozóki, **Á.** Mohácsi, **L.** Horváth **and G.** Szabó, *Sens. Actuators B*, 134, 2, 1027–1033 (2008).

[8] A. Pogány, Á. Mohácsi, L. Horváth, Z. Bozóki, G. Szabó, 2008: A photoacoustic system for measuring ammonia exchange between the biosphere and atmosphere, *Geophysical Research Abstracts,* 10, EGU2008-A-01665, 2008 SRef-ID: 1607-7962/gra/EGU2008-A-01665.

[9] A. Pogány, Á. Mohácsi, A. Varga, Z. Bozóki, Z. Galbács, L. Horváth and G. Szabó, *Environ. Sci. Technol.*, 43(3), 826–830 (2009).

[10] M. Norman, C. Spirig, V. Wolff, I. Trebs, C. Flechard, A. Wisthaler, R. Schnitzhofer, A. Hansel and A. Neftel, *Atmos. Chem. Phys. Discuss.*, 8, 19791–19818 (2008).

[11] E. Nemitz, C. Milford, M.A. Sutton, *Q. J. R. Meteorol. Soc.* 127, 815–833 (2001).

[12] L. Horváth, M. Asztalos, E. Führer, R. Mészáros and T. Weidinger, *Agr. Forest Meteorol.* 130, 282–298 (2005).

[13] M. Mauder, T. Foken, C. Bernhofer, R. Clement, J. Elbers, W. Eugster, T. Grünwald, B. Heusinkveld and O. Kolle, *Biogeosciences*, 5, 451–462 (2008).

[14] P. Schotanus, F.T.M. Nieuwstadt, H.A.R. de Bruin, *Bound.-Lay. Meteorol.*, 26, 1, 81–93 (1983).

[15] M. Mauder and T. Foken, 2004: Documentation and Instruction Manual of the Eddy Covariance Software Package TK2. UNIVERSITÄT BAYREUTH Abt. Mikrometeorologie, Arbeitsergebnisse Nr. 26. *http://www.bayceer.uni-bayreuth.de/mm/.*

[16] A.J. Dyer, *Bound.-Lay. Meteorol.*, 7, 363–372 (1974).

[17] T. Foken, *Angewandte Meteorologie*, Springer-Verlag Berlin Heidelberg 2006, pp. 325.

[18] T. Weidinger, J. Pinto and L. Horváth, *Meteorol. Z.*, 9, 3, 139–154 (2000).

[19] T. Weidinger, Á. Bordás, D.T. Mihailovic, A.Z. Gyöngyösi, A. Machon, K. Pintér and L. Horváth, 2007: Uncertainties in surface layer flux calculations using gradient and profile methods, *First Serbian Congress on Theoretical and Applied Mechanics,* Kopaonik, Serbia, April 10-13, 2007. Proceedings, edited by D. Sumarac and D. Kuzmanovic pp. 267–274.

[20] H.N. Lee, *J. Appl. Meteorol.,* 36, 1416–1423 (1997).

[21] L. Horváth and M.A. Sutton, *Atmos. Environ.,* 32, 3, 339–344 (1998).

[22] L. Horváth, H. Fagerdi, and M.A. Suttom, Long-term record (1981-2005) of Ammonia and Ammonium concentrations at K-puszta Hungary and the effect of Sulphur Dioxide emission change on measured and modeled concentrations. *Atmospheric Ammonia* edited by M.S. Sutton, S. Reis and S. Baker, Springer 2008, pp. 181–186.

[23] Machon, L. Horváth, B. Grosz, T. Weidinger, K. Pintér and Z. Tuba, 2008: *Measurtement and modelling of fluxes of nitrogen compounds above a semi-natural grassland ecosystem in Hungary,* Open Science Conference Reactive Nitrogen and the European Greenhouse Gas Balance. February 20th and 21st, 2008, Het Pand, Ghent, Belgium, edited by the NitroEurope IP Secretariat, pp. 65–66.

INDEX

B

C

D

E

F

G

H

I

N

O

P

T

U

V

W

X

Y

Z